T0275882

INDUSTRIAL PROCESS SCALE-UP

INDUSTRIAL PROCESS SCALE-UP

A Practical Innovation Guide from Idea to Commercial Implementation

Second Edition

JAN HARMSEN
Consultant, Harmsen Consultancy BV, Zuidplas, Netherlands

Elsevier
Radarweg 29, PO Box 211, 1000 AE Amsterdam, Netherlands
The Boulevard, Langford Lane, Kidlington, Oxford OX5 1GB, United Kingdom
50 Hampshire Street, 5th Floor, Cambridge, MA 02139, United States

Library of Congress Cataloging-in-Publication Data
A catalog record for this book is available from the Library of Congress

British Library Cataloguing-in-Publication Data
A catalogue record for this book is available from the British Library

ISBN: 978-0-444-64210-3

Publisher: Susan Dennis
Acquisition Editor: Anita Koch
Editorial Project Manager: Michael Lutz
Production Project Manager: Prem Kumar Kaliamoorthi
Cover Designer: Christian Bilbow

Typeset by SPi Global, India

Dedication

I dedicate this book to my wife Mineke. She inspires me endlessly and provides the outside view, as defined by Kahneman in Thinking—Fast and Slow.

Contents

Preface

I had a career of 33 years in Shell in fields of oil refining, bulk chemicals, biomass refining, biotechnology, and fine chemicals. I worked in exploratory research, process research, process development, process design, process start-up, and process de-bottlenecking; hence, in all process innovation stages.

In 2010, I started my consultancy company and advised many companies from various industry branches such as food, fine chemicals, bulk chemicals, biotechnology, ceramics, and fertiliser on sustainable process innovation.

In this career, I have seen many failures in each innovation step from idea to commercial implementation. I searched, many times, the literature to see what is available in information to prevent these failures from happening, but I found only a few articles describing only a few critical elements for successful process innovation and scale-up. This triggered me to write a book: Industrial Process Scale-up, which got published in 2013.

Now, after 5 years, I have written a revised second version of this book. I received several comments on the first version. Main critical points were: It was not suitable for pharmaceutical and fine chemical scale-up; and it did not treat scale-up from pilot plants, such as pilot plants available for testing for the new application, from technology providers. Those critical points have been taken care of in this book. Special sections deal with pharmaceuticals and fine chemicals. One chapter is dedicated to scale-up from pilot plants. I used the book for in-house courses and obtained feedback from the industrial participants. That is also incorporated in this version in combination with considering latest literature on process scale-up and innovation. Hence, this second version is indeed a revised version.

This book is focused on industrial process innovation and scale-up. It deals in detail with risk identification and risk reduction. It does now also include in-detail process equipment scale-up separately from integral process scale-up.

I think that this book will be of use in the process industries for anyone active in any part of the process research, development, design, start-up, or operation. This book is also suitable for industry courses as it provides the scale-up methodology systematically.

Preface

Acknowledgement

I gratefully thank the reviewers of the book for their comments. They were numerous, of high quality, and delivered in time. The reviewers are: Ben Bovendeerd of Technoforce, Michel Eppink of Synthon, Rene Bos of Shell and Ghent University, and Albert Verver of FrieslandCampina. So, the reviewers range from a technology provider with pilot plant facilities, a process researcher of a pharmaceutical company, a process developer from a bulk petrochemicals company, and a process developer from a food company.

CHAPTER ONE

Industrial scale-up content and context

1.1 Purpose and set-up

This book is mainly meant for industrial process innovators. The methods and guidelines provided for them in this book serve three purposes. The first purpose is to provide guidelines so that process innovation projects can be turned into successful commercial scale start-ups, rather than failures. The second purpose is to obtain the best process concept in terms of economics and other criteria, so that the new process is accepted by society and is competitive in the market. The third purpose is to provide guidelines to have innovation project executions that are lowest in cost and in elapsed time.

The need for this book is mainly based on the statistics that 50% of novel process introductions are disasters (Bakker et al., 2014). A disaster, here, is defined as having more than 30% cost growth beyond the budget and more than 38% schedule slippage. The statistics have been gathered from over 12,000 projects from all kinds of process branches by Independent Project Analysis (IPA), as reported by Bakker et al. (2014). Also, mega-scale projects often fail, as reported by Merrow (2011) and Lager (2012). Several projects had not reached design capacity even 5 years after the beginning of the start-up.

The effects of a commercial scale implementation failure for a company can be enormous. It is not only the additional capital investment needed and the revenue losses, but also the loss of trust of clients that the company faces regarding delivering their products as promised. It also means the loss of trust of top management in the innovation power of the company. This can strongly affect the budget for future process innovation projects. Also for technology providers, an implementation failure can have a large negative effect on future sales.

Industrial Process Scale-up
https://doi.org/10.1016/B978-0-444-64210-3.00001-9

The set-up of this book follows the stage-gate approach. The stage names are obtained from Harmsen (2018). For the first stage, discovery and concept methods and guidelines are presented that ensure that the best concept is identified and selected. In the subsequent stages, methods and guidelines are presented to reduce the risk of implementation to such a low level that start-up cost and start-up times stay within the budget.

Because part of the innovation project failures is due to taking shortcuts of the available guidelines, pitfall warnings are added at each stage-gate chapter. Throughout the book starting with Section 1.3.3, it explains and shows why project innovation shortcuts nearly always end in commercial disasters.

The nature of this book is prescriptive and not descriptive. The guidelines and methods have been proven or are plausible. These guidelines and methods can be used to generate essential design information, to assess risks and to mitigate risks to such a low level that commercial implementation is successful—and the innovation pathway is rapid and efficient. This book also provides real industrial innovation cases with additional learning points. The book is a description of an industrial best practice for scale-up. "An industrial practice is a cooperative human activity in which different professional disciplines work together to develop, produce and sell a product" (Verkerk et al., 2017). Major professional disciplines involved in specific innovation stages are, therefore, mentioned.

The book is intended for industrial process researchers and developers of process industry branches. Special attention in this book is given to pharmaceutical and fine chemical processes for each innovation stage. Furthermore, it will be of use for contract researchers and technology providers to see how and when they can interact with process industry manufacturers and engineering contractors.

The book, however, does not contain descriptions on how to manage and organise industrial research, development, design, and process engineering. It also does not contain detailed process design guidelines for the commercial scale detailed design. For that, the reader is suggested to refer other books on industrial management and process engineering such as Harmsen (2018), Bakker et al. (2014), Dal Pont (2011a, b), and Lager (2010).

1.2 Scale-up definition methodology and risks

1.2.1 Scale-up definition

Often, the term scale-up is used to simply state that a larger production capacity is employed, without any reference to whether this scale-up was successful and how this scale-up was achieved. To also include these elements of scale-up, we use the following definition in this book:

Process scale-up is generating knowledge to transfer ideas into successful commercial implementations.

Knowledge generation involves literature reading, consultation, experimenting, designing, and modelling. The main purpose of this knowledge generation is to be able to assess risks and to reduce risks to acceptable levels for the successful commercial scale implementation. The word "ideas" is stated in this definition rather than "concepts", as concept generation from ideas is also considered part of the scale-up. Also, the term "scale-up from laboratory scale" is avoided as laboratory experiments are also part of the scale-up knowledge generation.

Successful implementation means that the commercial scale process meets the design targets within the planned start-up time.

The purpose of industrial process scale-up is then mainly risk reduction needed for success. For people working in the process industries, this is a nearly trivial statement and Merrow's book on industrial megaprojects, based on more than 1000 industrial cases, proves that, indeed, in direct commercial implementation without proper industrial research and development, the risks of failure are always too high to take (Merrow, 2011).

For most academics, however, this statement is not trivial at all, because in the academic world the purpose of research is to generate understanding, knowledge, and theory. The word 'risk' does not enter in research papers about process innovation and is also not found in process innovation books. Jain et al. (2010) do not contain any description of a goal for innovation. Vogel (2005) and Betz (2011) only state that the goal of industrial research and development is to achieve competitive advantages.

1.2.2 Scale-up methodology

The scale-up methodology of this book is based on knowledge generation for risk identification, risk assessment, and risk reduction. Risk identification of a new process concept is already very difficult, because not all relevant information will be available. If a certain piece of information is not available, then it may be identified as an unknown. But for certain risks even that information may be lacking; I even don't know what I don't know. Table 1.1 shows these two different types of knowledge gaps, their associated risks and information plans to close the knowledge gaps.

Table 1.1 Types of knowledge gaps, risks, and knowledge generation plan

Knowledge gap type	Risk type	Knowledge generation plan
I know what I don't know	Specific and limited	Specific research
I don't know what I don't know	Unknown	Integrated process test

Risk identification is, therefore, carried out several times during the innovation project. Each time more information has been generated, more risks items will be identified and consequently risk assessment will improve. If the risks are too high, risk reduction plans will be made and executed in the next innovation stage. The risk dimensions envisaged are Safety, Health, Environmental, Economical, Technical, and Social (SHEETS criteria defined in Harmsen, 2018). The methodology focuses, furthermore, on guidelines and methods that are cost-effective and efficient. The effectiveness is obtained by providing guidelines on project target and constraints.

1.2.3 Stage-gate innovation method

Innovation is defined here as project management from idea to commercial implementation. The innovation effectiveness and efficiency are obtained by the stage-gate approach described by Harmsen (2018). This means that potential failures for the project are discovered as early as possible with little effort and, if they cannot be corrected, the project stops. For the structure of this book, we use the most elaborate stage nomenclature mostly used for the chemicals sector as a generic structure:

- Discovery
- Concept
- Feasibility
- Development
- Detailed design (including procurement and construction)
- Start-up

Other industry sectors such as food and Oil&Gas have different stage names, but the stage sequence and content are very similar. These are, therefore, easily treated in this book using the generic structure.

For the food sector, the stages are (Verver, 2018):

- *Orientation*: idea and concept generation
- *Creation*: process development: lab-scale, bench-scale, pilot plant
- *Preparation*: engineering, construction, commissioning start-up
- *Implementation*: commercial production

For the Oil&Gas sector, the stages are (Bos, 2014):

- Discovery
- Development
- Demonstration
- Deployment

For pharmaceuticals, the stages involved, obtained from Levin (2006) and Kane (2016), are:

- *Discovery stage*: The new molecule is assessed on its activity.
- *Pre-clinical stage*: The new active molecule (product) is defined.
- *Clinical phase I stage*: The new product and its formulation are made in a small-scale pilot plant for tests.
- *Clinical phase II stage*: The new product is made in intermediate scale pilot plant for further clinical testing.
- *Clinical phase III stage*: New product is made at larger scale plant for studies with many patients.
- *Approval stage*: The product is approved by the regulatory body and decision for commercial scale production is taken.
- *Manufacturing stage*: The new product is produced and sold.

Special attention is paid in each chapter to process innovation steps for all industry sectors and to process innovation for pharmaceutical industry. Section 4.7 treats how and when to consider process options for that industry sector.

Here is a birds-eye view of generic stages and their content. In the discovery stage, a proof of principle experiment is carried out at laboratory scale. In the concept stage, only information is generated for a feasible process concept design. In the feasibility stage, a commercial scale design and a down-scaled pilot plant design and costing are made. Depending on production scale and overall complexity of a respective industrial process, one can differentiate between a dedicated integrated pilot plant or piloting of individual unit operations. In the stage-gate, a decision to invest in the pilot plant, given the economic prospect of the commercial plant, is made. The implementation stage involves detailed engineering, procurement of equipment, construction, commissioning, and start-up.

For a de-bottlenecking project, it is advocated to also follow all innovation steps, rather than directly start with the end of the development step; defining the front-end loading for the next step. By starting with the discovery step, various de-bottlenecking options are considered. In the concept stage, the best options are defined and the best selected. In the feasibility stage, the best option is worked out in a design, evaluated, and all risks are considered, and in the development stage, some new aspects can be tested. These extra steps may take a few weeks to a few months, but will improve the quality of the de-bottlenecking project considerably.

In each subsequent stage, more information is generated, risks are more clearly identified, and more robustly mitigated to acceptable levels. If at any stage-gate the risks are estimated to be too high, or the cost of further development is higher than the final benefits of commercial operation, then the project is stopped, so that only a small amount of money is lost. In this way, innovation is not only effectively but also efficiently executed. This stage-gate approach facilitates, furthermore, communication about the status of the innovation to internal and external stakeholders and to external innovation partners.

1.2.4 Scale-up and design role

Making designs plays a role in each innovation stage. Making a design first reveals knowledge gaps, which leads to a plan to fill the gaps. Second, the design result is a communication means, as it shows what the innovation is about. For each innovation stage, a section on design will, therefore, be included. As the stages promote, the level of detail in the designs increases as well.

1.2.5 Scale-up behaviour and risks

This section provides some technical insight why process scale-up so easily goes wrong when critical success factors for scale-up in research, development, and design are not fulfilled. The insight is provided by the following aspects of new processes:

A: Chemical reactivity, including corrosion rates, can easily vary by a factor 10^9 by small changes in, for instance, water content, trace amounts of organic acids, halides (chorine), or metal ions. These trace components may be in the feeds to the process or formed in the process. The effect of these components can be rapid corrosion, foaming, and/or fouling, causing the process to fail. This behaviour may not show up in laboratory tests with pure feedstocks and short test durations.

B: The number of parameters in a process easily exceeds a 100. The combined behaviour of small changes in parameters often cannot be predicted well by models, e.g. due to their strong non-linear interactive behaviour and the lack of thermodynamic and physical data to support the computational models.

C: Dynamic time scales for components to build up in the process can be very long, in the order of months, in particular when recycle streams are involved. These build-ups will not show up in short duration laboratory scale tests.

D: The hydrodynamic behaviour and their effect on mass transfer, heat transfer, mixing, and residence time distribution often change with scale-up, causing a poor performance of reactors, heat exchangers, and separators.

E: The combined effect of A, B, C, and D cannot be predicted by models.

Aspects A–C are dealt with in Chapters 2 and 3. Aspects D and E are treated in Chapters 3–5.

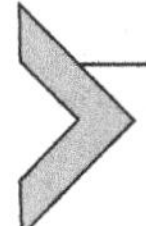

1.3 Process industry systems

1.3.1 Value and life cycle chains

Value chains are strings of intermediate product mass flow connections between companies and the final consumers. Each company adds economic value to the mass flows. These value chains from native feedstock to consumer products can be short and involve only a few branches such as in basic food products.

It can also be very long such as in consumer food products manufactured from multiple ingredients, such as infant formula, or in consumables (e.g. soap), where the steps involve crude oil refining, steam cracking, higher olefin conversion to alcohol, and then blending with fragrances and other additives, each with their own supply chain.

If, for instance, the crude oil feedstock is changed into a renewable feedstock, then in general this also means that new connections between industry branches must be negotiated. The same holds when a new product for a new market is developed. However, new product development is outside the scope of this book.

Innovation involving new partners takes, in general, considerable time as companies that hitherto had no contact and have their own vocabulary now must learn to communicate. The largest miscommunications occur when both use the same term, but mean totally different things about it.

An example from my own experience is of the term scale-up. In a joint research programme for a new process of a petrochemical company and a pharmaceutical precursor production company, we discussed the next step after the research. The petrochemical company considered the scale-up a big step involving many years. The pharmaceutical precursor company considered this a small step. After a long discussion, the petrochemical company said: "But scale-up involves designing and building a pilot plant", upon which the other company people said: "We already have the pilot plant". Then it became clear that both used the word "scale-up", but they meant differently.

The term life cycle is, in general, used in combination with the word analysis or assessment. In life cycle assessment (LCA), all process steps from native feedstock to destination such as waste incineration (called cradle-to-grave) or to end of cycle recycle and reuse (called cradle-to-cradle) are taken into account and also all mass inputs from nature and all mass outputs to nature. The differences between life cycles and value chains are that, firstly, life cycles are about all mass flows related to all steps, whereas value chains are about economically added values by each step and, secondly, that life cycle assessment is used to determine the total environmental impact of a product over the whole life cycle.

1.3.2 Industrial complexes

In industrial complexes, many processes are connected in many ways. Often, processes are owned by different companies. The complexes often contain a crude oil refinery, a steam cracker, producing olefins from a side stream of the oil refinery, and several chemical processes converting olefins to chemical intermediate products such as polymers, solvents, resins, and others. The processes are connected with many different streams to each other. In the Rotterdam industrial complex, for instance, an intermediate producer Huntsman is connected with 18 different streams to the other plants in the complex (Harmsen, 2010a).

1.3.3 Processes

The simplest definition of a process is a system of connected unit operations that converts a feedstock into a product.

For most applications, both feedstock and product have clear specifications and can be bought and sold on the market. However, many products must meet performance specifications, e.g. nutritional value of taste. For those products, a new process also means extensive product testing to ensure that the product is accepted by the market. Even for a new process for an existing specification product, some product testing by the clients will be needed. Specifications do not completely define a product. New trace components may be present in the product, for which no specification has yet been defined. The client may also have expectations from the existing product, which are not defined by the specification, such as described in Section 8.1.

If, also, the product is new, then product development is needed. How to execute a combination of product and process innovation is given in detail by Harmsen (2018).

1.3.4 Unit operations

Process technologies of all these industries have in common the factor that they are based on classic unit operations. Each process consists of one or more unit operations. Each unit operation has its own generic knowledge base of a combination of transport phenomena of mass and heat and momentum and their corresponding thermodynamics. In case of reactors, chemical conversion is added to these phenomena.

Unit operations based on fluid mechanics include fluid transport (such as pumping and pipe-flow), mixing/agitation, filtration, clarification, thickening or sedimentation, classification, and centrifugation. Operations based on heat transfer include heat exchange, condensation, evaporation, furnaces or kilns, drying, cooling towers, cooling and evaporative crystallisation, and freezing or thawing. Operations that are based on mass transfer include distillation, solvent extraction, leaching and absorption or desorption, adsorption, ion exchange, humidification or dehumidification, gaseous diffusion, crystallisation, and thermal diffusion. Operations that are based on mechanical principles include screening, solids handling, size reduction/grinding, flotation, filtration, and extrusion. Design methods for these unit operations can be found in handbooks such as Perry's Chemical Engineering Handbook.

For most commercial scale unit operations concept, design computer packages are available in so-called flow sheet computer programs. Scale-up of these units still, however, has risks if the unit operation has not been applied at commercial scale for that application and, in most cases, pilot plant development is required to validate the design methods applied and to identify the unknown-unknowns to the extent possible. Chapter 4 provides methods to decide on whether a pilot plant is needed.

1.3.5 Major process equipment

Each unit operation consists of a combination of major process equipment connected by pipes and flanges. A distillation unit operation, for instance, will consist of a column with internals, a heat exchanger at the top and the bottom. It may have pumps to circulate the fluid flows through the heat exchangers. These types of equipment are called major process equipment.

At scale-up, major process equipment will also increase in size. Reliable design and construction of the large scale is of utmost importance for successful implementation.

1.3.6 Dispersed system level

The dispersed system level is about bubbles in a liquid, catalyst particles in a reactor, and other dispersed phases in a continuous phase. Mass transfer and mixing are important phenomena at this system level. These phenomena are very important for the process performance and are, in general, scale-dependent (and application-dependent).

1.3.7 Chemistry level

The chemistry level is the smallest system scale of relevance to processes. At this level, chemical reactions are described. Often, these reactions are facilitated by a catalyst. The catalyst itself is consumed in this conversion, but enhances or moderates the various reaction velocities by changing the activation barriers. If the catalyst speeds up the desired reaction relative to the undesired reaction, it also increases the reaction selectivity. Catalysts typically are very sensitive to small changes in conditions such as temperature and to low concentrations of impurities, which may come from inputs to the process, from corrosion of construction materials, and from undesired reactions in hot areas such as distillation bottoms.

1.4 Partners and stakeholders for innovation

The following industry partners and stakeholders can be distinguished:

- Manufacturers
- Clients
- Suppliers
- Government
- Civilians
- Non-governmental organisations (NGO)
- Technology providers
- Engineering contractors
- Contract research organisations (CRO)
- Academia

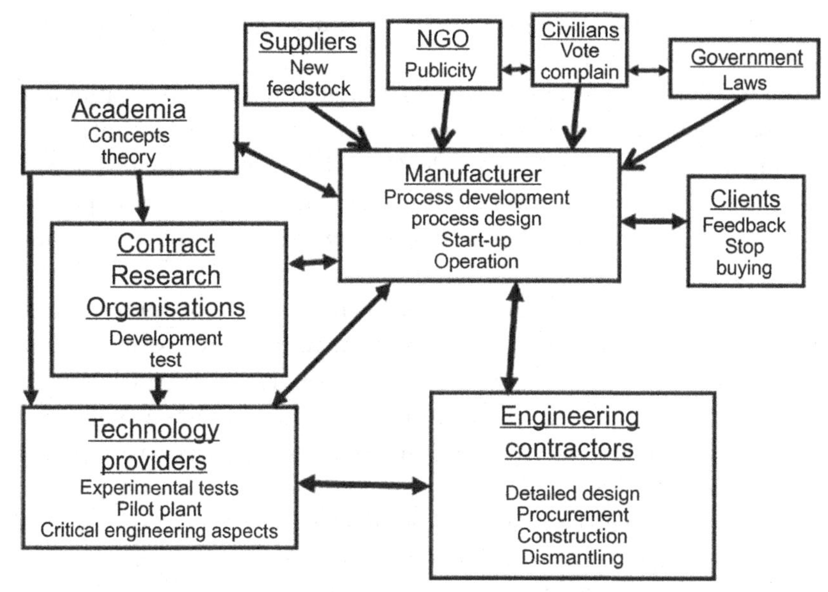

Fig. 1.1 Roles and relations innovation partners and stakeholders.

Fig. 1.1 shows their potential relations in process innovation. These partners and stakeholder groups will be shortly described in relation to process innovation.Manufacturers are companies that convert feedstocks from suppliers into products for their clients. Often, these companies have a research and development department to generate new processes. Also, employees directly involved in the manufacturing process are a source of process innovation ideas.

Manufacturers in the process industries are classified into branches.

The major branches are the following:

- Crude oil refining
- Metal ore refining
- Paper and pulp
- Bulk chemicals
- Fine and specialty chemicals
- Pharmaceuticals
- Food
- Agricultural products (feed)
- Consumables
- Ceramics

Each branch has its own characteristics in the process capacities employed, the way they operate, and the way research and development are carried out.

Due to these differences, technologies proven in a certain branch often fail when applied in a different branch.

Clients of manufacturers can be industrial companies or consumers. Industrial companies can initiate process innovation at the manufacturer by asking for a lower cost of the product, or a lower environmental impact of the product, or better product performance. They are also very important stakeholders in process innovation, as shown in the liquid-liquid extraction case of Chapter 8.

Suppliers of feedstocks are, in general, not sources of innovation at the manufacturer. But the manufacturer can initiate process innovation by asking a supplier for a lower cost or lower environmental impact. Innovations with a large total effect on cost and environment will more and more be carried out by collaborations over a larger part of the supply chain.

Government can play a role in process innovation by more stringent laws on safety, health, and environment and by subsidising process innovation.

Civilians living nearby the process can play a role in innovation asking for a safer, healthier, and environmentally friendlier process.

Non-governmental organisations (NGO) can be a source of innovation in the same way as civilians.

Technology providers can be very small innovative firms specialised in one novel process technology or larger firms with many innovative process technologies. Some have good relations with university groups, providing them with new ideas for innovations to valorise. They often have their process technologies protected by patents and other forms of intellectual property rights, such as copyrights on drawings and software. They provide the technologies to product manufacturers and to engineering contractors.

Engineering contractors for the process industries are often very large companies who often carry out complete EPC (Engineering, Procurement, Construction) process projects for manufacturers, which include process design, equipment procurement, and construction. They may have a process innovation department, but often they have relations with technology providers to generate process innovations.

Contract research organisations (CRO), in general, obtain process concepts from others, such as universities and manufacturers, and develop processes to the end of a pilot plant stage or to small-scale commercial implementations. The development effort may be paid by the manufacturer directly. The CRO can also develop the process at their own expense and then sell the technology (protected by patents) to manufacturers.

Academic research often generates radically novel process concepts, often on their own initiative. The concepts are often in the embryonic state. It needs others to convert these concepts into feasible solutions.

CHAPTER TWO

Discovery stage

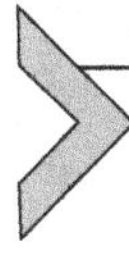

2.1 Obtaining process ideas

2.1.1 Obtaining ideas in general

In the discovery stage, it is very important to have many process ideas, so that the whole solution space is known, and the best process concept can be identified. In the discovery stage, it is also good to maintain a few promising process concepts for some time. Working on a few concepts can trigger additional ideas. It also means that, if one concept fails, another concept can be picked up. In subsequent innovation stages, the concept will be defined in more detail, but the fundamentals of the concept will, in general, not be challenged. As the final implemented process must be better than the competition, it is of utmost importance that the best concept is identified in this early stage.

The sources of process ideas are many. They may result from:

- Literature searches
- Visiting conferences
- Brainstorming and brainwriting
- Doodling by individual employees
- Studying existing processes
- Serendipity

Literature searches should not only be focused on scientific articles, but also on using general browsers and searching for websites, patents, and documents from companies and institutes.

Conferences are also an important source of ideas; not only from the presentations, but also from discussions in the corridors and breaks.

Brainstorming and brainwriting with a diverse group of people can generate useful ideas.

Doodling and Friday afternoon experiments of individual employees are also an important source of ideas.

For studying the existing process, I have a few suggestions derived from my own experience. I remember from when I was advising senior

Industrial Process Scale-up
https://doi.org/10.1016/B978-0-444-64210-3.00002-0

technologist in a department of advising technologists at a large chemical manufacturing site that young process engineers would go to the plant managers and ask them what problems they had, so that he or she could solve them, the plant managers said either: "We do not have problems, or plant runs fine", or they said: "Yes, we have a heat exchanger that rapidly fouls, solve that".

Later, the young engineer would then discover that it was nearly impossible to even analyse the nature of the fouling or the cause of the problem and that many advising technologists before him had been given the same problem and were also not able to solve it. The reason for this behaviour of plant personnel is that they know that, when the young trainee will start, he/she will ask them enormous amounts of questions and/or he/she will come up with a solution to the problem that will not work, and again, they must spend lots of time to prevent the implementation of the bad solution. May be my observation does not hold for other companies or does not hold anymore anyway.

But, still, here is a hint for obtaining a list of problems from existing plants. Offer the plant manager that you, as process trainee, are part of the operating shifts for 3 weeks and be part of all five shift groups. When in the shift groups you explain your purpose and listen to the operating personnel about what they know and think about the process, that traineeship will give you lots of insight in process problems. Still, typically, it will take 6–12 months before you are fully "trusted".

I also have a hint for the management of advising technologist departments: Have a mixture of experienced and young process engineers, so that the young engineers can express their problems freely and obtain advice how to best proceed.

Another source of ideas is when a conventional existing process requires additional capacity because of market demands. A de-bottlenecking analysis study should then first be performed to determine the bottleneck and so find the lowest cost solution for the capacity increase. The de-bottlenecking analysis study is another very good reason for visiting the existing plant and again be part of the shift operators for 3 weeks. If a process simulation model is available, then the actual process conditions and behaviour can be compared with the simulation model and the model can then, first, be improved and, secondly, be used to simulate the conditions and flow rates for the increased capacity. With the list of those conditions, the bottlenecks may be identified in discussion with the plant managers and other personnel.

2.1.2 Obtaining new process routes

New process routes to the same product should also be explored in the discovery stage. New process routes mean new chemical conversion pathways from the same feedstock to the product and new chemical pathways from alternative feedstocks. All kinds of catalysis (homogeneous, heterogeneous, enzymatic, microbial) should be considered. The number of different pathways from a given feedstock to a given product easily mounts to 10. Including different feedstocks, the number easily mounts to 100. Literature searches and brainstorming with chemists and biochemists are proven methods for obtaining new process routes.

As new process routes are a major source of break-through process improvements, this step should always be included in the innovation project.

2.2 Assessing process ideas

2.2.1 Assessing and ranking many ideas

In the discovery stage, many ideas will be needed. Generating a few hundred is common. Even if a carbon copy of a process will be designed, it is of value to consider new ideas. An example is provided in Chapter 8, Section 8.6, where a carbon copy was changed by incorporating a new process section for waste water treatment.

The many ideas must be ranked so that the most promising ideas are identified for further work. A good and useful method of ranking is by having a group of experienced innovators to, first, generate ranking criteria and, second, score the ideas against these ranking criteria. As most promising is linked to the innovation capabilities of the company and the values of the company, assessment should be carried out by employees of the company. A few persons from outside the company can be added to avoid narrow-mindedness. The criteria should cover following aspects:

- Fit with long-term company strategy
- Economic potential
- Feasible to develop by company (with outside contributions)
- Safety-Health-Environmental issues can be overcome

If the company is relatively small, with a small portfolio of innovation projects and a small budget, then following criterion can be applied in addition:

Development pathway and time-to-market.

The development pathway and time to market should be estimated with in-house experienced people. Projects for which they find it hard to make define the development pathway and the time to market should be scored low for this criterion.

Ranking of the ideas can be simply done by scoring the ideas against the criteria. This should be done individually by the panel members to avoid group bias. Then the scores should be collected, and averaged values and the value range should be determined. Finally, the average scores per criterion should be set and the overall score per idea obtained by multiplication. The advantages of obtaining the overall score by multiplication is that the ideas that score for all criteria reasonably well obtain a much higher score than ideas that score high on a few criteria but low on others. The highest scoring ideas will be more robust to uncertainties than the low scoring ideas for two reasons: (A) Defending the ideas for management will be easier if the ideas score well for all criteria as they have no obvious weak spots and (B) If in the development some weakness appears for a certain criterion, then the idea may still score reasonably well.

2.2.2 Technology readiness level method

A very good and efficient way of assessing an idea or process concept is using the Technology Readiness Level (TRL) (Table 2.1). It has been derived by

Table 2.1 Technology readiness level (TRL) definitions for process technologies

TRL level	Description	Innovation stage reached
TRL 0	Idea generated and explicitly stated	Discovery
TRL 1	Experimental proof of principle individual key novel process elements	Discovery
TRL 2	Process concept design provided	Concept
TRL 3	Proof feasibility process concept design by techno-economic assessment	Concept
TRL 4	Process experimentally validated by integrated mini-plant experiments	Feasibility
TRL 5	Process techno-economics assessed by professionals in process industry	Feasibility
TRL 6	Process technology demonstrated in industry by pilot plant	Development
TRL7	1st commercial scale plant in operation	Implementation
TRL 8	Learning points 1st plant incorporated in 2nd commercial design.	Deployment
TRL 9	2nd Commercial process operation meeting all specifications.	Deployment

Harmsen (2014) from other TRL tables and has been modified for this book to fit the innovation stages names used. It can be used for any technology reported in writing. It is to be applied by starting at TRL level 0. As soon as written information for the next level is there, that next level has been reached. The procedure is repeated until the information for a level is not there. The previous level is then the established TRL level for that technology.

In exercises with the method, it appeared that engineers and scientists nearly always scored a higher TRL level than the document provided warranted. It appeared that the engineers and scientists concluded that the next level was obtainable and used that conclusion to arrive at a too high TRL level.

Companies that want to use the TRL level table for technologies provided by external parties should ask for documents describing the technology developments and check the validity by visiting the experimental test facilities.

2.3 Process design sketch

2.3.1 Process concept design essentials

A process concept design will at least contain a block scheme with input and output flows. Preferably, the potential feedstock sources and product destinations are provided (Harmsen and Jonker, 2012). Sometimes the input sources and the output are not completely known, but only that the feedstocks and products are available on the market. In cases where the feedstock and/or product is not on the market, indications will be given as to how these feedstock sources and/or product destinations will be found. For instance, by searching for alliance partners who may be able to provide the feedstock or buy the new product.

The advantages of having a process design sketch in this very early discovery stage are that it helps to identify information needed, which helps to plan the concept (research) stage. It helps to communicate the idea and indicate the technical feasibility and economic advantages over the present conventional process. Fig. 2.1 shows the information flows of design and research steps.

Fig. 2.1 shows roles of process concept design in generating the information flows.

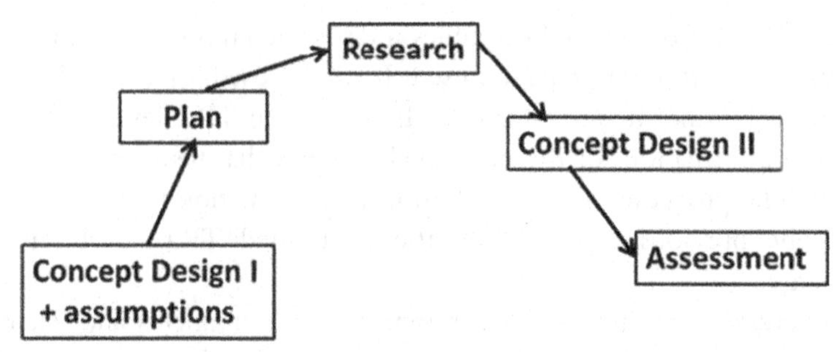

Fig. 2.1 Information generation steps research stage.

2.3.2 Reference case for comparison

A reference case is the best commercial scale process for the same product. The new idea must be better for several criteria and not worse for the other criteria. A very good description of this reference case and then a comparison of the new idea with the reference case on all criteria helps enormously in determining the present status of the idea and what needs to be done to turn the idea into a concept and finally into a commercial implementation that is far better than the reference case.

2.4 Proof of principle

Proof of principle are laboratory experiments to prove that the individual fundamentals of the new idea could work. The challenge for the researcher is, on the one hand, to keep the experiments simple and low cost, and on the other hand, make the result convincing for the budget provider to invest in further development of the idea. If it is a novel reaction, then an experiment showing that the reaction produces a significant amount of product may suffice. If it is novel separation technique, then showing experimentally that the product outputs have sufficient purity may suffice.

2.5 Discovery stage-gate evaluation

2.5.1 Discovery stage-gate purpose

At the gate between the discovery stage and the concept stage, a decision must be taken whether the process idea should be pursued or be terminated. In large companies, a decision will be taken on the whole portfolio of new process ideas. The ideas that are sufficiently promising will first be selected.

Then a ranking will be carried out and the top options will pass the gate. The total budget available for process research will determine how many process options will be researched.

The selection and ranking will be carried out by a panel of experienced research and development managers and business managers based on the information provided in reports.

2.5.2 Reporting discovery stage

The reports will, in general, contain the following items (van Harmsen and Eck, 2004):

- A reference case (best commercial scale process for same product)
- A rudimentary description of the process concept
- Proof of principle experimental evidence
- Strategic business fit of product
- Intellectual property position
- An economic evaluation
- A safety health environment assessment
- Major critical items to be researched
- Assessment of concept stage cost
- Outline of steps to commercialisation

The first two items have been described in Sections 2.3–2.5. For the other items, here are some guidelines.

A fit with the business strategy must be shown, if the idea is about a new product. If it is an idea of a new process for a product produced already by the company, then it is of value to check whether the company wants to extend the market. Some caution is, however, needed here. The business section may say that they have the long-term view that the market will not grow and/or also that the market share will not be increased. But this view can often and rapidly change. It is sometimes better to show the large potential of the new idea for product cost reduction and product quality increase to the business and not ask for a long-term business view.

The intellectual property position should also be indicated. It is tempting to assume that a rapid patent search will show whether the idea is novel or not. However, in practice, it is very difficult for a chemical engineer to assess novelty in the formal IP sense. If it seems to be novel, a thorough patent search should be part of the proposed research programme. If patents exist, then it should be reported when the patents were filed and who owns the patents. A patent attorney should be involved in an early stage if developing own IP position is considered. Novelty is a necessary but not sufficient requirement here. A "freedom of action" study should be part of the development plan. The timing of that can be very tricky, because they are always very country-specific, whereas in the early stages of the R&D one may not have concrete deployment plans. The research proposal should contain a plan on how to deal with those patents. Should an alliance be made with the owner or should the research be directed to avoid patent infringement? Are there other options to obtain freedom of action?

Also, the own intellectual property position and policy should be determined. If the company decides to generate patents or other means, such as design rights, copyright, or trademarks, to protect their innovations, then additional experiments may be required to cover a wider scope for the invention. For technology providers and contract research organisations, patent generation is often important to generate new business. For product manufacturers, patent protection may be important, but often free to apply their own inventions is even more important. Often, if it is decided not to pursue patents, the inventions are published to obtain that freedom of action.

A rudimentary economic evaluation, in general, helps to sell the research plan to obtain budget. Because only a proof of principle experiment is available, the economics can only be preliminary and rudimentary. For a novel (catalytic) reaction, an indication of a potential feedstock cost reduction compared to the reference case and existing process may be reported. For a novel separation, capital cost savings and/or energy savings relative to the reference case may be reported.

A preliminary safety, health, environmental (SHE) assessment will be made. A list of components and their SHE properties should be listed. A comparison with the reference case should be made, to show that the new idea has lower SHE risks than the reference case. If the risks are higher, then it will be hard to pursue the idea even if on other aspects, such as cost, the idea is beneficial.

All major critical items for success should be stated in combination with the research items to address these critical items and reduce their risk to an acceptable low level. Identification of these risks for the new idea is not trivial. Experienced process developers should be consulted to obtain the list of critical items.

Table 2.2 can be used as an additional checklist to identify critical items. A few risk items of that table will be explained here.

Table 2.2 Risk items and competence mapping new process

Risk item	Relevant?	Company competence	External competence
New market for product?			
New customers?			
New Country?			
New product?			
Product differs for existing customer?			
New feedstock?			
New supplier?			
New waste water stream?			
New waste streams?			
New safety aspects process			
Runaway reactions?			
Reaction enthalpy uncertain?			
Phase equilibria data uncertain?			
New health aspects?			
New environmental aspects?			
New social acceptance?			
New process?			
New technologies?			
Solids processing?			
New chemistry?			
New (modified) catalyst?			
Catalyst decay rate (including poisoning)?			
New recycle stream?			
Stream compositions unknown?			
Unit operation new to application?			
Equipment new to application?			
New construction material?			
Process control?			
New process employees?			
Mass transfer limitation?			
Mixing involved?			
Residence time distribution?			
Heat transfer?			
Intellectual Property and freedom of action?			
Patent position obtained?			
Competitive when implemented?			
Regulatory requirements (food, pharma), Food law requirements, Hygienic design, Cleanability of equipment and process?			

Product from new process is new to client

Often, the new process will be to produce an existing product. If the product is a performance product, then the risk that the product from the new process will perform differently will be easily identified and proper risk reduction measures taken by sending product samples from the pilot plant for the new process to the client. For specification products, however, manufacturers may assume that if the product from the new process is within the specification limits, the product will be okay for the customers. However, the new process may contain new trace components, which may affect the performance at the client. If then the product from commercial scale process appears not to be acceptable for the client, a large problem occurs. Also, for specification products, samples from the pilot plant of the new process should be sent to the client so that this problem is identified early and action can be taken. Section 8.1 describes an industrial case where a product sample of a specification product was sent to the client, meeting all specifications, but still the client could not accept the product.

New waste water stream

The new process may have a waste water stream to be sent to the biological waste water treatment facility. This waste water stream may contain trace components which kill the microorganisms of the waste water facility. The lowest cost solution often is to change the concept design and include a waste water unit operation that produces clean water that can be used inside the process site, for instance, as boiler feed water. The other output waste stream may be incinerated.

Runaway reaction behaviour

Runaway reaction behaviour is not always identified as a risk. Especially when the reaction is slow and easy to maintain in steady state in the laboratory test, process developers may neglect this risk. However, if a misoperation occurs, resulting in a higher reaction temperature, runaway conditions are easily reached. It is far better to perform runaway reaction experiments for all new reaction sections of the new process so that the risks can be assessed and mitigation actions in the early concept design can be taken.

Reaction enthalpy value uncertain

Presently, reaction enthalpies are obtained from flow sheet packages. The enthalpy values may be calculated using basic property models. They may be also obtained from databases, for which the enthalpy may have been determined at standard conditions. If, however, the reactions are at high temperature, then the reaction enthalpy can deviate from the calculated value. This may happen, in particular, for gas phase reactions where reaction mixture heat capacity may change from the standard conditions, and thus, affect the reaction enthalpy. It is, therefore, recommended to determine the reaction enthalpy at the reaction temperature.

Estimate of next (concept) stage time and cost

An estimate of the concept stage effort and timing involved can be made by using the critical items for success as starting points. In general, this estimation is too optimistic. It is, therefore, useful to have regular progress meeting in which the actual progress made is compared with the estimated timing. In review meetings with the management, improved estimates of effort and timing will be presented and agreement obtained.

Estimate of steps to commercial scale implementation

It is also very useful to outline the steps to commercialisation. Options for partnership later in the development can be stated and discussed and preparations can be started.

The report will be sent to the panel prior to the meeting. At the meeting, the basic idea and a summary of the critical items with the research plan should be presented, followed by a discussion, and finally, a decision should be made.

2.6 Pitfalls discovery stage

Independent Project Analysis has analysed over 12,000 process innovation projects. Fifty percent of these projects turn into disasters upon implementation. The analysis resulting in causes for these disasters is reported by Bakker et al. (2014). Harmsen (2018) reported additional causes observed by an experienced innovation manager. We rename them here as pitfalls. Here are then the pitfalls for the discovery stage:

Lack of adequate analysis of potential solutions

Lack of strategic fit

Lack of freedom to operate

From the previous sections, it will be clear that these pitfalls can be prevented by following the guidelines on generating solutions, on intellectual property, and on stage-gate evaluation criterion: strategic fit.

CHAPTER THREE

Concept stage

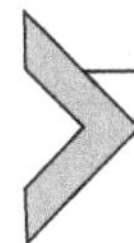

3.1 Detailed assessment of ideas

3.1.1 New process routes assessments

In the concept stage, the new process routes from the discovery stage are first assessed in some more detail and the best are selected. A rapid method for this assessment is based on economic potential of the process routes by considering the feedstock cost per ton of product and the number of process steps for each process route.

For the feedstock cost per ton of product, the feedstock cost per ton of feedstock and the yield of product on that feedstock are needed. The latter may be not available. An estimate can be obtained by assuming maximum theoretical yield derived from the stoichiometric equation.

For the number of process steps, the minimum number of process steps can be taken, also derived from the stoichiometric equation, and it is to be decided whether a catalyst and solvent must be separated from the intermediate product stream leaving the reactor. The lowest feedstock process route should be ranked highest. For nearly equal feedstock cost, the number of process steps should be used for a final ranking (Harmsen, 2018). In this way, the number of process routes for further investigation is reduced to one or two.

For pharmaceuticals and fine chemicals, chemical routes to the desired product are always explored. However, different process routes and various process media options such as:

- Not using solvents, but internal process streams
- Continuous processing with continuous reactors

are less explored. If this is done at the concept stage and continuous production is already decided and explored in this stage, then time to market can be reduced and often the reactive hold-up is reduced by a factor of 1000, such that safety precautions are easier defined and obtained.

Industrial Process Scale-up
https://doi.org/10.1016/B978-0-444-64210-3.00003-2

3.1.2 Detailed assessment best ideas

For the top ranked ideas from the discovery stage, a detailed assessment can be made. Table 3.1 may be of use by answering for each risk item whether it is relevant, whether the company has in-house competence for addressing the aspect, or whether external competence is obtainable. If the answers

Table 3.1 Risk assessment new process concept

Risk item	Relevant?	Competence in-house?	Competence external?
Intellectual property rights			
Patent position obtainable?			
Freedom of action?			
Competitive when implemented?			
Product composition different			
New feedstock			
Water feed to the process			
New waste water stream			
New waste streams			
New safety aspects process			
Runaway reactions			
New health aspects			
New environmental aspects			
New social acceptance			
New process			
New technologies			
Solids processing/handling			
New chemistry			
New (modified) catalysts			
Catalysts decay faster than design			
New recycle streams over most process units			
Stream compositions unknown			
Reaction enthalpy uncertain			
Unit operation new to application?			
Equipment new to application			
New construction material?			
Process control			
New process employees			
Mass transfer limitation			
Mixing involved?			
Residence time distribution			
Heat transfer tubular reactors			

Relevant? means: Should the risk item be addressed?

are all no, then the design must change such that this risk is no longer relevant. If that is not possible, then either the project should be abandoned or, if it is strategic, a long-term research program should be started to develop the competence.

Table 3.1 can also be used to transform tacit knowledge of experienced people into coded knowledge by having them answered the questions and asking them their specific experiences on which the answers are based.

3.1.3 Cooperation with external innovation partners

Using the competence gap analysis, the company may decide to cooperate with external innovation parties. The benefits are that knowledge is obtained and that the cost of obtaining that knowledge is in general lower than generating that knowledge by in-house research. For radically novel process concepts requiring new knowledge fields to be developed, cooperation with external partners should be considered.

There are, however, also risks by involving external parties. These are:

A: Miscommunication, for instance, because the same words, such as scale-up, research, and pilot plant have very different meanings in the two companies.

B: Lengthy contract negotiations before and during the research and development slowing down the innovation rate.

C: Loss of intellectual property and intellectual property rights.

Therefore, the following steps are advised before entering into a full joint venture on innovation:

Step 1: Capabilities needed for the innovation project for successful implementation are not available in-house. Then ask the question: Can the concept design be modified so that these external capabilities are no longer needed? If so, then a decision between the two alternative designs should be made.

Step 2: If external knowledge is needed, then it should be explored whether the information can be purchased, for instance, by contract research with a contract research organisation with clear clauses on intellectual property disclosure and rights.

Step 3: If Steps 1 and 2 are not sufficient and a collaboration on research and development is considered, then it is good to know that in my experiences collaborations between product development manufacturers are often not successful, due to engrained differences in culture to do research and development and business in general, causing continuous misunderstandings and even fights and stalemates.

Collaborations between process industry manufacturers and well-established technology providers, however, are in general successful, because they have clear and different business purposes and turf fighting hardly occurs.

Step 4: If Steps 1–3 are not successful, then a pre-competitive open innovation platform can be joined, where a broad technology field is explored in which the specific own process fits in to some extent. The own process will then in general be not disclosed, but the broader knowledge generated in the open innovation platform can be used to build up the necessary knowledge and capability to solve the specific problems of the specific process.

3.2 Identifying potential showstoppers

Determining potential project showstoppers of a new process is a very important part of the concept stage. Because if it appears that these potential showstoppers appear to be real by performing experiments, then the process innovation project can be stopped and wastage of large amounts of money and research capacity is prevented. Often, the process concept is modified in such a way that the showstoppers are removed. If the incentives of the modified concept are still sufficient, then the project can be redefined and continued. Here is a hint to design experiments to identify showstoppers. Think in the term: "Fail fast". It is a mixture of a negative and a positive word. I may overcome the psychological hurdle of researchers to want success and not failure. A successful fail fast experiment can then be celebrated.

Here are some methods to determine potential showstoppers:

a. Make a process concept design containing all input streams and all output streams and make sure that it contains all process steps connecting all input streams to all output streams (Douglas, 1988; Harmsen and Jonker, 2012; Seider et al., 2010)
b. Interview experienced process designers and process developers of the company and ask them what they think are potential showstoppers of the process concept

In making the process concept design, many assumptions will have to be made, because several pieces of information will not be available yet. The assumptions which are key to the success of the process concept are all potential showstoppers.

Experienced process designers have a lot of tacit knowledge that becomes active when showing the process concept design. They will highlight many areas of concern and may also state: This will never work. Their knowledge also concerns the capability of the own company to develop and design the process concept. So, this knowledge will also reveal whether the own company can develop this process.

Perhaps, a word of warning would not be missed to invite external academics to judge the feasibility of the concept. Often, they will state that the concept is not feasible and then give one or more reasons why the concept is not feasible. The best way of dealing with this advice is to write down all the reasons why it will not work and then test these reasons experimentally at a later stage in the company. What, in general, is not fruitful is to start a discussion on the reasons why it cannot work.

3.3 Proof of concept

The experimental test of the potential showstopper of the new concept should be done in such a way that it is convincing to the experimenter and his/her management that the new idea is based on either a real physical and/or a chemical phenomenon or not. This experiment is often called proof of principle experiment. Because in the early research stage little money is available, it will take ingenuity to execute the experiment with little cost. Companies with a prominent innovation strategy often let researchers spend 10% of their time and budget on these types of experiments without writing a proposal upfront. This allows the researcher to feel free to carry out experiments with a low chance of success, but potentially with a high commercial impact.

An example of a proof of concept is provided in Section 8.6, where a novel distillation set-up was tested in a small glass laboratory set-up revealing a foaming problem.

3.4 Concept design

3.4.1 Purpose concept design

Concept design in the concept stage serves several purposes. Firstly, break-through process technologies are generated by making concept designs. The basis new principle will have been generated and tested in the discovery stage. Exploring the basic idea by sketching new process concepts will help enormously to obtain a simple, safe, and low-cost process design.

Secondly, by designing process concepts, gaps in information will appear. At first instance, these gaps will appear as assumptions the designer has to make to continue his design effort. In second instance, these assumptions must be validated by research. So, concept design helps in identifying what needs to be researched.

Thirdly, concept design facilitates an assessment on safety, health, environment, and economics, as the design provides information on what chemicals are used at what conditions, sizes and compositions of input and output streams, and defined required process steps.

Fourthly, the design is a communication means between the research team and the management. It shows what the innovation object is about.

Section 8.10 shows what can happen if a concept design is not made in the concept stage.

3.4.2 Concept design

Classic process concept design (also called process synthesis) methods based on unit operations are provided by Douglas (1988) and Seider et al. (2010). Douglas starts from input and outputs and then these are connected by unit operations. Seider starts from reaction followed by separation, conditions and phase change setting, and finally tank integration. So, Douglas works from the outside to the inside and Seider from the inside to the outside. Both have advantages. Experienced industrial process concept designers do not stick to either of the two. They often start from the output side. Then do the reaction and separation as a block flow solution and then evaluate the input in connection with an optional feed separation section and then rearrange the block flows and then decide on the best unit operation for each block and decide on recycle flows. After that, the process is simulated and optimised. Additional design methods such as industrial symbiosis may be used to fit the process design in the context. Harmsen (2018) contains systematic process concept design as well as many special design approaches to obtain a good concept design.

3.4.3 Break-through concepts by function integration

If a break-through design is desired that is lowest in investment cost, operation cost, maintenance cost, and is most reliable, then process intensification by functional integration should be applied. It starts with identifying essential functions. These functions are best put as blocks in block flow diagrams in which arrows represent mass flows. Take the essential blocks and combine the functions by changing the process conditions such that they get into the same temperature and pressure window. Then sketch a process in which these functions are in the same equipment.

Most common functions in the process industries, mentioned by Harmsen (2018), are:
- Mass transport
- Mixing
- Heat transfer
- Reaction
- Separation
- Shaping

The equipment in which the functions are combined can be distillation columns in which, for instance, reaction and separation are combined into reactive distillation and in which gas flows up are driven by pressure drop and liquid mass flows down are driven by gravity. Mixing and reaction can, for instance, be combined in a pump or in an extruder. Other examples of function integration are dividing wall columns in which two distillation functions are combined in one column, a continuous crystalliser in which crystal growth/shaping/classifying and heat transfer by evaporation are combined.

The advantage of this type of process intensification is that the number of equipment reduces enormously. This then means lower investment cost, lower energy cost, lower maintenance cost, and reduction of failure risks. This latter point of having less equipment and thereby lower risk of failure is emphasised by Table 3.2, showing failure statistics of process equipment. By having less equipment means less valves and flanges and pumps, with high failure rates.

Table 3.2 Equipment failure in a percentage of plants where this occurred (Merrow, 1988)

Equipment type	Plants failure %	Comment
Pumps	31	In most cases seal failure
Valves	28	Incomplete shut off
Dryers	19	Failure due to material shows slightly different behaviour
Compressors	17	
Agitators	17	
Conveyers	14	Failure due to material is different from originally intended

Total percentage failure is >100% because several plants had more than one type of equipment failure.

3.4.4 Process modelling and simulation

Process modelling and simulation helps to optimise the process concept and to generate the stream sizing and compositions and in sizing the main equipment. For complex process phenomena inside certain equipments, specific Computational Fluid Dynamic (CFD) model building and simulation helps to optimise the equipment shape and details of the configuration. These models also help in the next stages to predict the behaviour at large scale, and so, can be used for scale-up. Table 3.3 shows these two major modelling types used in process concept design with their purposes and limitations.

Many process flow simulation packages, also called flow sheeters, are now available on the market. They have libraries for unit operations and for predicting physical properties. They have also optimisation routines so that, for a defined design target and design constraints, they can determine optimum design parameters. Some can also do dynamic simulations.

The limitations of these packages are, firstly, when a novel process step concept involving combinations of functions in a novel way is researched, and then the concept cannot directly be simulated with the package. When knowledge of the mixing, residence time distribution, mass transfer, and/or heat transfer is becoming available, then by smart combinations of existing library models sometimes still the process simulation package can be used.

Custom-made models for these new process sections often can be made by a modelling department inside the company or by an outside modelling contractor company.

Model predictions using flow sheeters are very sensitive to the values of the physical properties such as phase equilibrium. This equilibrium can be predicted with special modules of the simulation packages. However, if the mixtures contain polar components and/or components that form hydrogen bonds, then these model predictions can be quite wrong. If phase equilibrium for these types of mixtures is predicted by the modules, then model validation with dedicated phase equilibrium experiments will be needed.

Table 3.3 Types of process modelling simulation packages

Simulation packages	Purpose	Limitation
Process flow sheet simulation	Sizing, optimisation, stream compositions	Not for novel equipment concepts. Not for multi-phase turbulent behaviour
Computational fluid dynamics	Simulate fluid flow, geometry optimisation	Not for gas-liquid and liquid-liquid bubble or droplet size prediction

Also, the whole simulation model must be validated using integrated pilot plant tests with the same process sections and recycle flows, before it can be used to design the commercial process.

Models also must include kinetics for reactions and crystallisation. For the latter, crust formation on the heat exchange surface must also be included in the modelling. If fouling occurs, then also this should be modelled.

Computational fluid dynamics (CFD) packages can now be obtained from vendors to predict complex flow behaviour often in combination with complex geometries. For one-phase flow, these packages are in general very suitable.

For multi-phase flow and in particular for gas-liquid and liquid-liquid phase systems, however, the CFD models are in general very unreliable for predicting bubble and droplet sizes and coalesce with mixing and break-up. This is caused by two different phenomena. Firstly, the real chaotic turbulent behaviour is modelled in the CFD using approximations. For the large-scale eddies, the model predictions are in general reliable. However, for the small-scale eddies, near the vicinity of a flexible surface the models are inaccurate. Secondly, bubble break-up and bubble coalescence are very complex phenomena, where often trace amounts of components present in the two-phase system show a preference for the interface and then either enhance coalescence or prevent coalescence. Because of this, the model predictions for bubble and droplet size are often totally wrong, and hence, also the mass transfer rates that often partly determine the net chemical conversion rates. For the same reason, CFD models for emulsion formation often are completely wrong.

This field is extensively researched by many academic groups for over 30 years, but the progress is slow. For a comprehensive review of the present status of two-phase flow simulations, the readers are referred to a review article by van den Akker (2010).

If experiments with the real phase mixture have been carried out, then model predictions can be validated. But if the experiments have been carried out for a small scale and the CFD model is used to predict the large commercial scale behaviour, still the model predictions can be wrong. For instance, because in the large scale the time scales for high intense turbulence areas where bubble break-up occurs and the low-intensity turbulence where bubble coalescence occurs are different and the "kinetics" of bubble break-up and coalescence are different from the small scale, where the turbulent field was more homogeneous.

3.4.5 Essential properties determination

Physical and chemical properties are needed to perform process concept synthesis and process concept design analysis. Most of these scale-independent properties can be found in all kinds of literature sources.

However, for some process steps such as distillation or extraction, accurate values of phase equilibrium are needed to determine whether distillation will be feasible for the desired separation performance. If the phase equilibrium properties are inaccurate, then it may be concluded that distillation is feasible, while in reality an azeotropic mixture may be formed, so that a one-column distillation is not a feasible option. If this is found out at the pilot plant stage, then a redesign is costly in time and money.

Another important property is the chemical equilibrium property for reactions. If the physical property prediction wrongly predicts that the reaction equilibrium is far to the product side and a pilot plant reactor shows that the equilibrium is much more to the feedstock side, then the whole process design must change.

A third important property is the heat of reaction. If the reaction is strongly exothermic or endothermic, a special heat exchange reaction system must be designed. Also, the small laboratory scale reactor may already suffer from strong temperature increase or drop effects, due to the high value for the heat of reaction.

A fourth important scale-independent data set is the chemistry of the reaction. This involves, firstly, the stoichiometry of the reaction. This means, simply stated, what molecules react to which molecules and how many of molecules are needed to form the desired molecule. Also, the stoichiometry of the major by-product reactions is needed, to make a first estimate of the yield of product on feedstock.

Finally, the reaction kinetics need to be determined to some extent. For the reactor-type selection, it is needed to know the reaction scheme; if it is a consecutive or a parallel reaction for by-product formation. And if it is a parallel reaction, then what is the relative order of the main reaction to the by-product reaction. Also, the relative activation energies are key for a rational reactor selection as it determines both the optimal level of the temperature and the optimal profile for *endo-* or exothermic reactions.

Some indication of the time required to reach the desired conversion is, in general, enough in the discovery stage to determine the reactor size. In the research state, accurate kinetic rate expressions are desired to allow reactor design optimisation. So, accurate physical property knowledge is needed in the research stage for certain process steps and for certain physical properties to allow a pilot plant design that has a reasonably high rate of success. If the new process design involves new components, then in most cases certain physical properties, critical to success, will have to be determined experimentally. This can be done inside the company or outside by institutes specialised in physical property determination.

3.4.6 Concept stage reporting

Table 3.4 shows a comprehensive list of reporting elements to complete the concept stage. This reporting can then be used for the stage-gate decision and when the gate is passed as the start for the feasibility stage.

Table 3.4 Concept stage-gate report elements

Content item	Description
Project design name	Unique name to design project, to avoid confusion in communications
Project goal	Clear statement of project goal
Safety, health, environment, economic, social constraints	Major constraints are listed
Required product output	Products quality specifications, production rate, market/clients envisaged
Other outputs	By-products and effluents specifications and potential destinations
Required inputs	Input specifications and rates of all streams and utilities and list of potential external and internal company suppliers. Raw materials extensive descriptions
Economics	Feedstock cost and sales revenues. Capital cost and energy relative to reference case or crude estimates
Safety base	Runaway behaviour laboratory experimental results toxicity data
Health base	Health hazards components data
Environmental base	All streams from and to the environment over the life cycle specified
Social acceptance	Social acceptance information of process and products
Research quality base	Properties quality, design accuracy, mini-plant results
Process description	Process block flow diagram with all inputs/outputs, technologies, special conditions

Continued

Table 3.4 Concept stage-gate report elements—cont'd

Content item	Description
	(temperature, pressure). Operation: batch/continuous
Novel elements	List of novel process elements with feasibility statement commercial scale
Strategic fit	Innovation type and projected effort fit to the company strategy
Intellectual Property status	Freedom of action, patent position
Feasibility stage cost	Cost estimate of executing feasibility stage
Innovation plan	Main items in next stages
Assessment	Benefits and Risks: Economic, safety, health, environmental, social and technical and regulatory issues; product safety status (e.g. Novel Food)

3.5 Concept stage-gate evaluation

The purpose of the concept stage-gate evaluation is to decide to continue to the feasibility stage or not. As the feasibility stage involves a reasonable design effort of a commercial scale concept design and a detailed pilot plant design and may involve some feasibility testing, this is an important decision involving a significant budget. The potential benefits should be sufficiently large to warrant this spending. Using the concept stage result report, a decision by management will be taken to let the project pass this stage-gate and provide budget for the feasibility stage.

3.6 Pitfalls concept stage

From IPA analysis as reported by Bakker et al. (2014) in Table 1.2 and from Pinto also reported by Bakker et al. (2014), the following pitfalls for the concept stage are obtained:

Lack of adequate analysis of potential solutions

Let new ideas starve to death through inertia

The first pitfall can be avoided by applying the guidelines in this chapter on analysing potential solutions. The second pitfall can be avoided by having a stage-gate assessment as soon as the concept report is available and by deciding on the budget for the feasibility stage at the same stage-gate meeting.

CHAPTER FOUR

Feasibility stage

4.1 Purpose of feasibility stage

The feasibility stage is a relatively new stage in the stage-gate innovation method. In the past, it was part of the development stage. The purpose of the feasibility stage is to determine the feasibility of the whole innovation project, considering all benefits and all risks. So that, at the end of the feasibility stage, the question can be answered: The large spending on the subsequent development stage is warranted in view of future benefits and risks? This feasibility stage is also called: Building the Business Case, or Opportunity Framing.

The chapter is structured as follows. First process equipment scale-up is treated in Section 4.2. Then, integral process scale-up by design and experimental validation is treated. The next sections are about specific decisions for a mini-plant and an integral pilot plant. Then, in Section 4.7, feasibility items specific to pharmaceuticals and fine chemicals are treated. The final sections are on concurrent development and commercial scale design, stage-gate evaluation, and pitfalls.

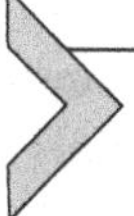

4.2 Process equipment scale-up

4.2.1 Introduction to process equipment type selection

A major decision to be taken is the choice of major equipment type for each process step in relation to reliable scale-up to commercial scale. This subject is not treated in process design textbooks. Harmsen (2018) treats the subject to some extent by discussing scale-up strategy and scale-up methods in relation to process equipment. His book has a table showing 10 process equipment types and the process equipment phenomena: residence time distribution, micro-mixing and mass transfer, and which scale-up methods are available (Harmsen, 2018). Here, we will try to explain how the choice for each process step can be made for a reliable scale-up strategy.

Industrial Process Scale-up
https://doi.org/10.1016/B978-0-444-64210-3.00004-4

The scale-up choice for process equipment, three elements interacting with each other, should be considered. The elements are:

Process equipment phenomena
Scale-up methods
Process equipment types

4.2.2 Critical performance phenomena

4.2.2.1 Overview of critical performance phenomena

Critical performance phenomena characterise the equipment for their process performance behaviour. These phenomena are also scale-dependent. For chemical reactors, some of these critical performance phenomena such as residence time distribution are well-described. In fact, more than half a century ago, this realisation has led to a new discipline: chemical reaction engineering. In this discipline, many of these critical performance phenomena concepts have been developed. That many of these phenomena are also relevant to many other process unit operations and process equipment is hardly known. I, therefore, made Table 4.1 showing the critical performance phenomena, their relevance to unit operations, and some simple concepts. The relevance of each phenomenon is described in subsequent sections.

4.2.2.2 Feed distribution

Feed distribution is the phenomenon that the feed entering the process equipment must be uniformly distributed over its cross-sectional area. At small scale, this uniform distribution is often easily obtained by single point feed in equipment with a small diameter. Often, the phenomenon at small scale is not even considered to be important.

For most concept designs, the concept of a perfectly uniform feed distribution is sufficient. But, for the feasibility stage, commercial design and, for the down-scaled pilot plant, such a simple assumption is no longer valid as the performance for most equipment strongly depends on proper feed distribution.

Upon scale-up, the phenomenon is nearly always very important for the process equipment performance. If no information has been gathered in the concept stage on the effect of feed distribution parameters on the performance, then only the brute force scale-up method (see Section 4.2.3) is available for reliable scale-up.

The feed distributor to be used for the commercial scale should be tested in a cold-flow test facility or information from other industrial scale applications should be obtained about its feed distribution quality.

Table 4.1 Critical performance phenomena

Process equipment phenomenon	Relevant to	Simple concept cases
Feed distribution	Reactors, distillation	Perfect uniform
Residence time distribution	Reactors, separators, extractors, ion-exchange resin beds, crystallisers, falling film evaporators, grinders, heat exchangers	Plug flow, ideally back-mixed flow, Fed-Batch feeding, axially dispersed flow, external recycle
Mixing (no, micro, meso, macro)	Reactors, mixers, extractors	Segregated flow (no mixing) Infinitely fast early mixed
Shear rate distribution	2-Phase contactors, centrifugal separators, extruders, grinders	Uniform
Mass transfer	Multi-phase reactors, distillation, falling film evaporators, extraction, adsorption, membranes, spray dryers, crystallisation	Not limiting step, totally limiting step
Heat transfer	Heat exchangers, reactors, crystallisers, distillation	Heat transfer not limiting (perfectly isothermal); totally limiting step
Impulse transfer	Spray dryers, fluid beds, and fixed beds	Negligible

The distributor of specific process equipment is commonly part of the proprietary design of equipment manufacturers, or technology suppliers, and needs to be tested for suitability in combination with a specific application.

For multi-phase systems, feed distribution is of enormous importance. For a bubble flow column, for instance, the gas feed distributor can determine the mass transfer performance. Also, the scale dependency of the effect of the feed distribution can be enormous.

To illustrate the importance of paying attention to feed distribution for the commercial scale design, here is a case from my experience. I have seen a commercial scale fixed bed reactor in trickle-flow, which after start-up showed very poor selectivity compared to the pilot plant. The technology provider could not find the cause of this deviation and paid the penalty fee for not meeting the performance specification on selectivity. Then, the problem was sent to me. I asked for the gas-liquid distribution design details.

With that, I went to our trickle-bed detailed design expert, who concluded that the number of liquid distribution points per square meter was far too low. Upon installing the correct distributor, the reactor reached its specification selectivity.

4.2.2.3 Residence time distribution

Residence time distribution (r.t.d) is about the phenomenon that some of the fluid parts entering the equipment have a shorter residence time than the mathematical mean residence time and other fluid parts a longer one. So, the fluid residence time has a distribution. The causes of these residence time distributions are many. Both geometry and flow regime play a role. For flow through a straight pipe in laminar flow, for instance, the fluid particles near the wall will have a lower velocity and thereby a longer residence time than fluid particles in the centre, while in turbulent flow a nearly uniform r.t.d. will be obtained.

A simple r.t.d. concept case is (ideal) plug flow, meaning that the residence time of each fluid part is the same. The fluid flows as a plug through the equipment. Another well-known (idealised) concept is that of "fully back-mixed" concept. Here, all fluid elements entering the reactor are instantaneously perfectly mixed with the fluid elements already in the vessel. This implies an exponentially decaying residence time distribution curve. Residence time distribution theory, and how it can be determined for any piece of equipment, by simple means, can be found in reaction engineering text books.

Residence time distribution (r.t.d.) for the concept design of reactors is always considered. But for separators, crystallisers, extractors, grinders, and heat exchangers, it should also be considered, as their performance in terms of conversion, selectivity, and product quality depends on it. For the feasibility stage design, this phenomenon should certainly be considered for each process equipment and the equipment with the most desired r.t.d. for the required performance should be selected.

A common misunderstanding is that fully back-mixed flow is an extreme of wide residence time distributions. This is totally wrong. A residence time distribution can be far more extremely distributed than a back-mixed flow. For instance, a considerable fraction can show a very short-cutting flow, causing a fraction of the flow with a very short residence time. For a reactor, this can mean less deep conversion than expected from back-mixed flow. The industrial scale-up case 8.4 describes this case.

Another common misunderstanding is that if, for a single-phase system, one knows the residence time distribution and the reaction kinetics, the conversion is fully determined. This is only true for first-order reactions. A back-mixed reactor in series with a plug flow reactor will have the same residence time distribution as that same plug flow reactor followed by that same back-mixed reactor, but, e.g. the conversion for a second-order reactor will be higher in the latter case.

A word of warning follows on the use of the term CSTR. This term has in reaction engineering the same meaning as back-mixed. However, chemistry students often use the term for a mechanically stirred tank reactor. If that reactor is used batch-wise, then it is in reality a plug flow reactor, so the opposite in residence time distribution of a CSTR.

Reliable scale-up methods are described in Section 4.3 and, for some process equipment, additional information is provided in Section 4.2.5.

4.2.2.4 Mixing

Mixing phenomena are important for the performance of reactors, mixers, and extractors. There are four degrees of mixing: segregated flow (absence of mixing), micro-mixing, *meso*-mixing, and macro-mixing. This is an enormous field of knowledge and hard to summarise in a few sentences here. The reader should investigate literature for applying the knowledge to his own case. We will treat here two simple extremes, segregated flow and micro-mixing.

Segregated flow

The absence of mixing is called segregated flow. An important application area for the latter concept of absence of mixing is ore conversion. The reaction takes place inside the particles, so mixing is absent. Often, deep ore conversion is desired. Due to the segregated flow, this asks for a r.t.d. of near plug flow even if the reaction order is zero. Westerterp et al. (1984) provides mathematical expressions for segregated flow, general r.t.d. expressions, and reaction kinetics. Calculations with those expressions show that an r.t.d. with a short residence time fraction of the total flow will cause a low conversion, regardless of the average residence time.

Micro-mixing

Micro-mixing is about the intimate mixing of reactants down to the molecular scale. Three cases can be considered, infinitely fast micro-mixing, slow micro-mixing, and complete segregation.

For the case of infinitely fast micro-mixing, the mixing rate is so fast that no significant reaction has taken place during this mixing. Using kinetics, and micro-mixing rates of the process equipment of the commercial scale, a quick estimate can be made whether micro-mixing is relevant for the case at hand.

Bovendeerd (2012) has provided a nice overview of micro-mixing aspects compared to macro-mixing for mechanically stirred vessels.

Macro-mixing	Micro-mixing
• Mixing on macro scale/scale of reactor	• Contact and mixing on molecular scale
• Governed by mechanical stirring	• Dominated by way of molecular diffusion
• Homogeneity by bulk transport of chemicals	• Enhance encountering of species
• Residence time distribution and backmixing	• Local effect of bulk turbulence
• Largest scale reduction of concentration fluctuations	• Smallest turbulent Eddies
• Macro-mixing time/blending time	• Micro-mixing time/ diffusion time
• Order of magnitude: tens of seconds to minutes	• Order of magnitude: milliseconds

If micro-mixing is slow compared to the reaction rate, then the micro-mixing model by Zwietering (1984) is recommended to explore what could happen at commercial scale. It is based on an easy-to-understand model of a jet starting from the feed, which mixes with fluid with bulk fluid composition with a mixing in flow rate related to the micro-mixing time. By-product formation and conversion during this micro-mixing time can be calculated with the simple model.

Computational Fluid Dynamics (CFD) can also be used to model the effect of micro-mixing on the reaction selectivity and conversion. Then, also, the reactor design can be optimised to minimise by-product formation for the commercial scale. That model can then be validated in the development stage in a pilot plant.

In the literature, the effect of micro-mixing on reactor selectivity is always described with complex reaction schemes. But micro-mixing effects can also occur for simple reaction schemes. In my career, I came across a reaction scheme where feed component A formed a fast equilibrium split reaction with component B and both components reacted modestly fast with component C to form products D and E. The commercial scale mechanically stirred reactor had a different D/E ratio from the small-scale lab-scale reactor, which could not be understood. I used the Zwietering (1984) model to calculate the D/E ratios and showed that the D/E ratio difference could be understood by differences in micro-mixing rates of the two scale. With the Zwietering model, conditions in terms of temperature and residence time could then be easily found to influence the D/E ratio.

Here are some rules of thump for considering micro-mixing effects. If the reactions are fast, with reaction times of less than 1 min, then it should be given further consideration. If then the reaction system also has a concentration order effect on the selectivity, then micro-mixing should be further investigated. For instance, by using the simple Zwietering (1984) model to get an impression of micro-mixing effects on the selectivity.

Micro-mixing and residence time distribution interaction

It is to be noted that, for segregated flow, process equipment performance can still be sensitive to residence time distribution in an unexpected way. For instance, for segregated flow with a zero-order single reactant, the reaction will be sensitive to the residence time distribution of the reactor, while for the instantly fast micro-reaction the reactor performance for zero-order reactions will be not sensitive to the residence time distribution, see for details and models for instance (Westerterp et al., 1984).

4.2.2.5 Shear rate distribution

Shear rate is the velocity gradient dv/dx, in which v is the fluid velocity and x the scalar space dimension. Most equipment have a nonuniform shear rate distribution. In mechanically stirred vessels, for instance, the shear rate is very high near the impeller blades, but further away the shear rate is much lower.

Shear rate in process equipment is applied to form fine particles (grinding) to create droplets or bubbles, or to create mixing. As the shear rate is often not uniformly distributed in the process equipment and this distribution is sensitive to the process scale and details of the dimensions, reliable scale-up is hard to achieve.

For single-phase systems, Computational Fluid Dynamics (CFD) is a powerful method to optimise the process equipment and, after validation, is also powerful to design the commercial scale process equipment reliably.

For processes that are sensitive to shear rate distribution, it is very important to consider the scale-up information for the various process equipment types under consideration. Reliable scale-up knowledge and a reliable scale-up method should be the overriding selection criteria.

To illustrate this point, I take emulsion polymerisation as an example. Emulsion polymerisation is sensitive to shear rate distribution, as the droplet size distribution is determined by the shear rate distribution. The droplet size distribution in turn determines the polymerisation quality and the final product particle size distribution. To the best of my knowledge, most emulsion polymerisations are carried out in mechanically stirred tank reactors, for which upon scale-up the shear rate distribution changes and thereby the product quality. By using the empirical scale-up method in which the recipe is modified to adjust for the scale-up effects, the desired quality is maintained. Choosing a different process equipment type, for instance static mixer, and applying the brute force scale-up method (increasing the static mixer diameter only and keeping all other parameters the same) would result in a far more reliable scale-up not needing recipe modifications.

A good textbook on this subject is van den Akker and Mudde (2014).

4.2.2.6 Mass transfer

Mass transfer per unit volume is defined as component transfer across an interface such as a gas-liquid interface of a dispersed bubble in a continuous liquid phase. It is determined by three elements, the concentration difference between the bulk of the fluids on one side and the other side of the interface, the interface specific area (a), and the mass transfer coefficient (k). The specific interface area (a) is inversely proportional to the bubble diameter. The mass transfer coefficient (k_l) at the bubble outside is affected by shear rate near the bubble. The mass transfer coefficient at the inside (k_g) is mainly affected by the bubble diameter, but also by the turbulence inside the bubble. The shear rate then determines to great extent the mass transfer of liquid-liquid, gas-liquid systems, and gas-solids systems.

For many process equipment types, mass transfer information is available often in the form of the Sherwood number (Sh) as a function of Reynolds (Re) and Schmidt (Sc). These correlations have often been determined for physical model systems such as water and nitrogen or air. The correlations have often also been determined in small laboratory equipment. The applications for commercial scale process equipment can be outside the confidence intervals of the correlations. Also, the commercial scale equipment may also have residence time distributions for the gas and liquid phases deviating from plug flow. Often, these aspects are neglected in the use of correlations. For the specific application, confidence intervals of the correlations should be checked.

In scale-up for processes where mass transfer is a key governing factor, the process equipment choice is very important. Process equipment for which reliable scale-up methods are known should be applied. A simple rule of thumb is provided here. If mass transfer is between a fluid and a solid phase, then mass transfer correlations, once validated experimentally, can be used reliably for commercial scale design.

For two-fluid systems such as gas-liquid and liquid-liquid, the large-scale mass transfer is often hard to predict reliably, due to break-up and coalescence behaviour of bubbles and droplets sensitive to local turbulent shear rates, which in turn are sensitive to scale. Moreover, these phenomena are also sensitive to trace components with interface activity, so small changes in feed compositions at commercial scale can change the mass transfer performance. To reduce the scale-up risk of the process, engineer can resort to process equipment where the mass transfer is not the limiting step, meaning that the mass transfer performance is so good that a change in that performance will not affect the process performance. Section 4.2.3 treats this subject in some detail.

A good textbook on mass transfer theory is by van den Akker and Mudde (2014). A very accessible and eye-opening book on multi-component mass transfer is by Wesselingh and Krishna (1990). They also highlight the pitfalls of simply assuming Fickian diffusion for multi-components systems and advocate the Maxwell-Stefan approach.

4.2.2.7 Heat transfer

Heat transfer is the heat flux from the bulk of medium one to the bulk of medium two across an interface. Heat transfer per unit volume is determined by the temperature difference between the two media, the heat transfer coefficient, and the specific interface area.

Heat transfer is an important phenomenon in reactors, distillations, crystallisers, and heat exchangers. In scale-up for processes where heat transfer is an important element such as in strongly exothermic and endothermic reactors, the choice of process equipment type is very important. An industrial case is the Shell gas-to-liquid hydrocarbon synthesis process step, described in Section 4.2.6. There the multi-tubular fixed bed reactor was chosen over the gas-sparged slurry reactor. Because of reliable scale-up, although the achievable product yield/selectivity was lower, the investment cost was higher.

The simple concept case names for heat transfer in reactors may need some explanation. If the heat transfer rate is not the limiting step, it means that the reaction heat is exchanged with a negligible temperature difference. This is called an isothermal reactor. If the heat transfer is totally limiting, then the temperature profile is totally governed by the heat transfer rate. A classic is example is the steam-methane reformer (SMR) reactor in which the endothermic reaction is so fast that the temperature profile is governed by the heat transfer performance of the furnace with its reaction tubes.

4.2.2.8 Impulse transfer

Impulse transfer is the transfer of impulse from the flowing process media onto the process equipment. The simplest version is fluid flowing through equipment. The force of the fluid on the equipment is directly observed by the pressure drop of the fluid from inlet to outlet.

Impulse transfer phenomena are important to select and size pumps and compressors for the process. They are also important for the integrity of the process at large scale. The pressure drop of the fluid flowing through a fixed bed, for instance, will cause force on the bed particles. At commercial scale, this pressure drop is often much higher than that at the laboratory or pilot plant test scale. The higher force on the particles may crush them, causing blockage and a process failure.

Another example of impulse transfer and scale-up effects is for bubble columns. By impulse transfer, gas bubbles cause a liquid circulation flow, which in turn causes an impulse transfer on the vessel wall and internals. These forces can be considerable and may break down the gas sparger and other internals.

Impulse transfer for the commercial scale can be predicted from fluid dynamic theory. The effects on the process performance via other phenomena, however, can in general not be predicted.

van den Akker and Mudde (2014) is a good textbook on impulse transfer theory.

4.2.3 Process equipment scale-up methods

4.2.3.1 Overview process equipment scale-up methods

Major proven scale-up methods for individual process equipment, nowadays used in industry, are:

- Brute force
- Model-based
- Empirical
- Empirical-model hybrid

Other scale-up methods:

- Dimensionless numbers
- Direct

Each process equipment scale-up method is described in the next sections.

It is noted for clarity that scale-up and scale-down of integrated processes are treated in Sections 4.3–4.7.

4.2.3.2 Brute force scale-up and scale-down method

In the brute force scale-up and scale-down method, the process equipment is a scale-down version of the commercial scale design in such a way that all critical factors and conditions for scale-up are kept the same. In practice, only for a few specific cases this can be done, because typically at least 1 or 2 key factors will inevitably change upon scale-down. One such example is a multi-tube process equipment where, upon scale-up, the number of tubes is increased, but the flow rate and tube dimension are kept the same; this method is also called the numbering up method. It is a very reliable scale-up method, although one should not forget the feed distribution phenomenon. It can be easily communicated to managers and it needs a limited amount of research, as it needs a limited amount of basic design data.

4.2.3.3 Model-based scale-up method

In model-based scale-up, the effects of scale-up on the unit operation performance are predicted by numerical models. The models contain all physical, chemical, thermo-dynamical, and hydrodynamical effects for the performance. The models have, of course, to be validated by means of experimental data.

This validation should occur on the interaction between the hydrodynamics (RTD, mass transfer, heat transfer, impulse transfer) on the one hand and on the reaction chemistry of the model, as in the case of a reactor performance, on the other hand. The pilot plant experiments should be such that not only a single point validation is carried out, but that experiments are conducted in which critical effects to scale-up such as fluid velocities are varied. Thus, the model predicted effects are validated with the experimental results. Hence, the design of the pilot plant must be such that it is possible to investigate the complete scaled-down operating window at its limits, for generating proper scale-up data.

The size of the pilot plant is a compromise between cost and meaningfulness of the obtainable data with respect to the model validation. The pilot plant experiments should be such that not only a single point validation is carried out, but that experiments are conducted in which critical effects to scale-up such as fluid velocities are varied; so that the model-predicted effects are validated with the experimental results.

The effects of much larger dimensions of the commercial scale than the pilot plant scale on the hydrodynamic performance behaviour can be validated in cold-flow "models", sometimes also called mock-up models. The residence time distribution (RTD), mass transfer, mixing rate, and heat transfer rate of the model can be validated with experiments.

If gas-liquid mass transfer is important, then the bubble size, bubble break-up, and/or coalescence are also important. In the hot pilot plant with the real gas-liquid mixture, the bubble size may be derived from the model validation and an indication whether the system is a rapid coalescing or a slow coalescing system may be obtained. The gas-liquid model system should then also be of the same nature. This can be obtained by using distilled water or salty water as the liquid.

This model-based scale-up method is often employed in the bulk chemical industry, for large-scale reactors in which the reactor dimensions affect the hydrodynamic behaviour via the reaction conversion and selectivity. Examples are bubble flow reactors and fluidised bed reactors—with or without internals—where the feed mixing rate and/or the RTD is critical to the reaction conversion and selectivity. CFD models are then validated with large-scale cold-flow models operated with model gases and liquids at ambient temperature and pressure.

4.2.3.4 Empirical scale-up

In the empirical scale-up method, the unit operation is carried out at several scales, often 3 or 4 scales. Often, the test units at various scales are given specific names such as bench scale, micro-reactor, mini-plant, pilot plant, business development plant, and demonstration plant. At each scale, the performance of the unit is measured and often certain parameters, such as stirrer speed, residence time, pH, feed ratios, chemical additives, temperature, and pressure, are adjusted to obtain the desired performance.

By plotting the required adjustments (to obtain the desired performance) to the experimental scale, a graph may be obtained revealing a systematic trend. Using this graph, extrapolations are made to the commercial scale. Often, scale-up trend effects of similar processes, including the effect of the commercial scale, are also available in the company. If the observed trends are similar to these previously obtained trends of similar processes, then some scale-up confidence is obtained.

Statistical models combined with Design of Experiments methods are useful in empirical scale-up, as with these methods maximum information is abstracted from the expensive experiments.

Besides being costly and time-consuming, the empirical scale-up method is not very reliable, as the underlying phenomena causing the scale effects are unknown and remain unknown. Also, the real critical scale-up parameters are unknown. But if the brute force method cannot be applied and the model-based approach cannot be employed, because the interactions between the hydrodynamics and the chemical reactions are very complex, then this is the only remaining method. Otherwise, the commercial scale design must be chanced into a design where the brute force scale-up method can be applied.

The empirical scale-up method is often employed for polymerisations and fermentations in mechanically stirred vessels. In most cases, the reactors are batch or fed-batch-operated.

For polymerisations, micro-reactors instead of the empirical method with mechanically stirred vessels can speed up the research and development enormously, because several development steps are not needed anymore. Section 4.7.3 presents an industrial case on this subject.

4.2.3.5 Hybrid model-empirical scale-up

There is a hybrid version of the empirical scale-up method in which modelling and simulation are carried out to interpret the empirical results and simulate and optimise the next scale-up step. In the next step, the empirical results are then used to validate or adjust the model. Again, the improved model is used to design the next step. In this way, the chances of success are increased.

The method is often used for mechanically stirred tank reactors for emulsion polymerisation. Often, the process is operated batch-wise. Model descriptions are available for reaction kinetics, but a reliable model for the droplet size distribution is in most cases not available. As the droplet size distribution also determines the final particle distribution and the latter is important for the performance in the process of the client, the scale-up is cautious multi-step empirical in which the kinetic model is validated for molecular weight distribution.

4.2.3.6 Dimensionless numbers scale-up

In the dimensionless numbers scale-up method, firstly, all dimensionless numbers are defined from a dimension analysis of all critical parameters, and second, the values of the dimensionless numbers are kept the same at scale-up from the test unit to the commercial scale. Zlokarnik (2002) provides a description of the method. For scale-down, the dimensionless number values of the commercial scale are kept the same for the down-scaled test unit.

This method, already known in the 1960s of last century, has not been used in industry, as far as I know. The first reason is that one cannot be certain that all critical phenomena are identified and captured in the dimensionless numbers. Secondly, no validation method can be defined to test that the dimensionless numbers are indeed capturing all phenomena. Thirdly, communicating the scale-up method to relevant stakeholders in the company, with no formal chemical engineering education, is nearly impossible.

The reason for treating this scale-up method in this book is that it is sometimes brought up in a company by someone and then others feel uncertain what to say. The above arguments may help to clarify the discussion.

4.2.3.7 Direct scale-up

Direct scale-up means directly design, construct, and start-up a novel commercial scale process without prior research and development work. This method is sometimes used by start-up companies for processes containing one major piece of equipment. Lack of experience and/or knowledge about process scale-up causes them to take this decision.

According to Merrow (2011), this method always fails. Also, technology providers of special process equipment sometimes take this decision for a new process application. Also, there it fails, as witnessed for instance in the Shell gas-to-liquids process for Bintulu in Malaysia (van Helvoort et al., 2014, pp. 230–267); where Shell experienced that certain equipment—not core to Shell's GTL technology itself—in hindsight appeared to be prototypes rather than proven technologies. The stream compositions to be processed were never tested for a vendor package of a subsection, including rotating equipment, valves, and adsorbentia. This caused corrosion and sealing problems (van Helvoort et al., 2014, p. 255).

Buying novel equipment or processes from technology providers always should be checked for (lack of) scale-up and design knowledge of the technology providers. The chance of success for direct scale-up is too small to pursue. It is far better to follow one of the proven scale-up methods provided in this chapter.

4.2.4 Scale-up characteristics of major process equipment types

Fixed bed multi-tubular heat exchange reactor scale-up

The multi-tubular fixed bed reactor consists of tubes filled with catalyst particles, which is cooled (or heated) via the tube wall. The tube outside volume contains the cooling fluid. In case of process temperatures in the range of 200°C–300°C, often the cooling is established by boiling water, with high coolant side heat transfer coefficients and that steam is directly generated. This is then used elsewhere in the process for heating up. This reactor type is suitable for moderately fast reactions, reaction times typically in the range of 1–100 s, for which the optimum particle size is larger than 0.5 mm, requiring heat exchange and near plug flow behaviour.

If all or many process equipment phenomena are critical to the commercial scale process performance and not all kinetics, reaction enthalpies, and the heat transfer coefficient are known, then brute force scale-up method is the only reliable scale-up method. Care should still be taken to ensure that the fluid distribution over the tubes is even, and that the catalyst is uniformly loaded of the tubes and with the same packing density. Specialist firms with experience with the catalyst shape and size and tubular sizing should be involved in the actual loading of the commercial scale reactors.

Unloading of the spent catalyst is also not trivial. The catalyst particles may stick to each other and/or to the tube walls. Specialist firms should be called upon to perform the unloading. Sometimes, air-lancing from the top is applied, for which specific procedures must be followed to obey safety, health, and environmental laws and regulations.

If all basic design data are known and validated, then the model-based scale-up method can also be applied. Correlations are available for fluid flow residence time distribution and mass transfer, even for trickle-flow operations (Gierman, 1988). Heat transfer coefficient correlations for tubular packed beds with small tube to particle diameter ratios ($<<10$) are, in general, not very reliable. This is because the packing near the wall is irregular, the local porosity in the radial direction locally ranges from 1 to 0.3, and the average porosity is strongly dependent on the mode of loading. It is recommended to determine the heat transfer coefficient, or better still the effective bed conductivity and wall heat transfer coefficient of the actual packing in the tubes experimentally in the pilot plant. This also holds for non-standard shapes of catalysts, including trilobes, quadrulobes, and even Raschig rings, for which the correlations in literature are not derived. For the simple overall heat transfer coefficients, this can be done, for instance, by feeding an inert gas (nitrogen) through the reactor and heating the reaction medium via the tube wall by the cooling fluid. Determining the effective bed conductivity and wall heat transfer coefficient is much more cumbersome and fuller of experimental pitfalls, so only recommended if proper expertise is available.

Bubble columns (with heat exchange)

In bubble columns, gas is fed via a sparger to the liquid inside the column. The rising bubbles cause the liquid to circulate. This circulation increases with vessel scale, affecting the bubble hold-up and thereby the gas-liquid mass transfer. Martin Obligado has experimentally determined this circulation rate with column diameters of 0.15, 0.4, 1.0, and 3.0 m. It appears that the liquid circulation flow rate increases in proportional with the diameter to the power 2.5. He also determined mass transfer coefficients for his model systems as a function of column diameter. The complete experimental work is available (Raimundo, 2015).

If heat exchange tubes are placed inside the vessel, then reasonably high heat transfer can be obtained by the circulating liquid. This process equipment is suitable for reactions where some back-mixing is okay for the reaction conversion (of both liquid and gas component) and selectivity and mass transfer limitations are negligible. The back-mixing can be reduced by having a so-called horizontal bubble column. The horizontal length is much larger than the (vertical) diameter and, especially, design baffles are placed in the horizontal liquid flowing through the vessel. A special baffle geometry is needed to ensure that no short-cutting flow occurs, as shown by Harmsen and Rots (2009).

In the absence of significant mass transfer limitations, reliable scale-up can be obtained using model-based scale-up with model validated in a pilot plant. Section 8.4 described a successful scale-up with such a horizontal bubble column reactor.

Gas-sparged Slurry Reactor with Heat exchange Scale-up

The gas-sparged reactor is also called a three-phase bubble column reactor. Gas is fed via a sparger to the liquid-solids slurry. The rising bubbles cause the slurry to circulate around the heat exchange tube, so that high heat transfer coefficients are reasonably obtained. This reactor is suitable for reactions where back-mixing is acceptable for the reaction conversion (of gas and liquid components) and selectivity, where mass transfers of gas to liquid and of liquid to catalyst particles are not the limiting steps, and/or where shear rate distribution does not cause particle attrition. It has the advantage that spent catalyst can be withdrawn and fresh catalyst fed, under normal operation. Some staging of the liquid phase can be obtained by having a so-called horizontal bubble column with a horizontal length much larger than the vertical height and with special designed baffles perpendicular to the liquid flow to ensure staging. Reliable baffle design for this staging is described by Harmsen and Rots (2009).

Separation of the catalyst from the crude product flow is the critical step for this process equipment. It is recommended to do this separation outside the reactor in a dedicated solids separation step.

A suitable scale-up method is model-based scale-up. This can only be done if the design is robust to uncertainty ranges for mass transfer coefficients, residence time distribution, and heat transfer coefficients. This, in turn, can be done by making a design, firstly, such that gas-liquid mass transfer is not the limiting step, secondly, by having sufficient liquid flow staging by a proved baffle design, and thirdly, using conservative heat transfer coefficient estimates. A reaction engineering design model based on kinetics, residence time distribution, and heat transfer can then be validated in the down-scaled pilot plant.

4.2.5 Process equipment type selection in view of reliable large-scale performance

In the selection of process equipment, three criteria are important:

- Performance is sufficient for the required function and specification
- Scale-up to commercial scale is reliable
- Lowest total cost over the life process time

The selected equipment must be able to perform the required function and meets the design specifications. Selection for this criterion can be done by selecting equipment for which the performance can be calculated, and experimental validation is feasible. Sections 4.2.2 and 4.2.3 are useful for this selection.

Scale-up to commercial scale can be done reliably. To this end, an analysis should be made of critical scale-up phenomena relevant for the performance in the equipment. Previous sections of this chapter should be helpful in this respect.

The total lowest fixed cost can be taken here as the main criterion for equipment choice. This holds if the process equipment options have the same product yield on feedstock, and in general, for the same performance specifications. This approach will not hold for the reactor/reaction section of the plant, as then also yield, and hence feedstock cost, is affected by the reactor choice and should be taken into account. The total fixed cost can be estimated from the investment cost, operating cost, and maintenance cost. Vendor quotes or internal cost data may be available for each process equipment on investment, operating, and maintenance cost.

In selecting process equipment using these three criteria, a trade-off must be made. If the commercial scale is large, then the reliable scale-up criterion will be overriding the lowest total cost criterion. For, if the commercial scale process does not produce the product, all will be lost. The next section will describe such a decision.

4.2.6 Historic case process equipment type selection Shell gas-to-liquid development

Helvoort describes in detail the historic development of the Shell gas-to-liquid (GTL) process (van Helvoort et al., 2014). Being a professional science historian, he writes in detail about the long and winding development road. At some moment in the development of the hydrocarbon synthesis, two options are considered for the reactor: A slurry reactor and a multi-tubular reactor. The slurry reactor is chosen as it can be operated for a long time without interruption for catalyst replacement and the investment cost is lower, due to higher heat transfer coefficients for the tubes inside the gas-sparged slurry bubble column reactor. Also, theoretically, a slurry type catalyst could achieve higher selectivities and catalyst mass-based productivities. Catalyst testing and further development is carried for some years. Then, the project moves to a different business section (Gas & Power), who wants to explore remote gas sources. The new director wants a technology that can be implemented at commercial scale as soon as possible. For the slurry reactor, scale-up is considered riskier and more time-consuming, requiring an intermediate demonstration step of 100 bbl/d, while the fixed bed multi-tubular reactor scale-up to commercial scale is considered feasible. As the catalyst life time has increased by the development effort considerably, fixed bed multi-tubular reactor operating in trickle-flow is now also considered feasible for long process runs (years).

The case shows that first process research and development is not a straight road, which can be precisely planned. Business environments can change, for instance, by changes in oil price and EPC contractors market price. Strategic changes of the company can affect the innovation project. Secondly, a choice made in the concept stage with the available information is revisited in the feasibility stage, with new business criteria, and more information are available. Thirdly, the case shows that reliable scale-up is a very important criterion to decide between different process equipment options with different scale-up uncertainties.

4.3 Process scale-up by design

4.3.1 Purpose commercial scale design in feasibility stage

The purpose of commercial scale design in the feasibility stage is to determine the feasibility of the new process concept in terms of safety, health, environment, economics, technical, and sustainability (SHEETS) (Harmsen, 2018).

This information will be needed for the feasibility stage-gate for deciding to make the investment for the development stage or not.

This design is also used to make a down-scaled pilot plant or mini-plant design.

4.3.2 Design choices in view of development cost

The commercial scale design for the feasibility stage will, in general, be a Front-End Loading (FEL-1) type design. It will contain all major equipments specified in type and size and construction materials, all stream sizes and compositions, and the instrumentation and control philosophy. Guidelines for such a commercial scale design are provided in text books such as by Dal Pont (2011a, b).

With that information, the investment cost, the potential gross income, and the return on investment are determined. Safety, Health, and Environmental criteria are applied by the defined stream compositions, equipment types and sizes, and emission streams. The technical feasibility is estimated from a check whether design methods and scale-up methods for the commercial scale process design are available for each unit operation.

By making the commercial scale design, a list of assumptions will also be made for missing information. This design knowledge lack may be solved by planning for the development stage to generate that knowledge by external consultation and/or experimental tests.

Alternatively, a different design is made for which all information is available and scale-up is reliable. The higher investment cost of this modified design is then traded off with the reduced development cost for not needing to determine the information.

An example of this type of reasoning is the gas-to-liquids process of Shell. For the synthesis reactors, two technology options were considered. One option was a three-phase fluidised bed—or slurry reactor—option. The other option was a multi-tubular reactor. The fluid bed has a higher heat transfer coefficient, hence needs less heat transfer equipment and allows for smaller catalyst particles with no mass transfer limitation; hence, smaller reactor volume and potentially higher selectivity. So, this reactor type has a lower investment cost than the multi-tubular reactor. However, the fluidised bed has a higher back-mixed behaviour, by which a higher fraction of heavy components will be formed, which can cause catalyst particle agglomeration, followed by deteriorating reactor performance. Because of this uncertainty for scale-up, the decision was taken for the multi-tubular reactor technology (van Helvoort et al., 2014, p. 183).

4.3.3 Design base

The design base should be reliable and explicitly described in a document. Firstly, the document should have a section describing the following point:

The new elements of the process. An element is new when it is used for the first time for this application, even if the technology itself has been used in other applications.

All essential research and development for scale-up as executed and reported should be taken into account for the design base. Then, the design base should be explicitly described. Table 4.2 is a content list of the design base with a brief description. It is derived from Bakker et al. (2014) for front-end-development (FED-1) and adapted for process development and extended by the author's knowledge obtained by my lifelong experience in the oil & gas and bulk chemicals process industries.

Table 4.2 Design base content FED-1 commercial scale process

Content item	Description
Project name	Unique name for project to avoid confusion in communications
Business goal	Strategic fit, product definition, capacity, and required economics
Design objective	Base for feasibility assessment and down-scaled pilot plant
Feeds (raw material) specifications	All process inputs are specified in composition and source
Product specifications	Product specifications and performance requirements for clients
Process context	Available sources for process inputs and destinations process outputs
Safety, health environment, social, economics	Constraints for design
Economics	Incentives commercial scale taking all cost into account Commercial scale investment cost estimate +/− 40%
Risk assessment safety	Operators, product users, local society of process, Initial HAZOP
Risk assessment explosion	Runaway behaviour reaction systems and mitigation measures
Risk assessment health	Health of operators, product users, local society of process

Table 4.2 Design base content FED-1 commercial scale process—cont'd

Content item	Description
Risk assessment environment	Environmental risks of process
Risk assessment economics	Economic uncertainties market size, sales price, sales volume in time
Risk assessment social	Social acceptance of process
Risks assessment technical	TRL assessment
Scale-up methods	Decision pilot plant, scale-up methods unit operations defined
Development cost and time	Estimate of development cost and time
Lessons learned	Lessons learned from previous scale-up projects for similar processes
Technology review	Process options considered and choices made
External knowledge	Which knowledge will be obtained from external parties
Process description input	Process design concept stage, block-flow diagrams, stream compositions, temperature, pressure ranges, operation mode (batch, continuous), physical properties, reaction enthalpies
Process description output	All major process equipment types and sizes, temperature and pressure, all stream compositions, and all heat flows

4.3.4 Purchasing a complete process technology

Sometimes, a complete process technology is purchased. The technology readiness level of Section 2.2.2 is then to be applied. If the process has been pilot-planted, according to the selling party, it is important to check in detail the pilot plant testing carried out. To that end, Table 4.3 has been made. It is a checklist for purchasing a novel process technology. It is based on the author's experiences and no claim is made that this checklist is complete.

If the process to be purchased also includes a detailed process design from an engineering procurement construction (EPC) company, then checking the quality of that design, and its corresponding documents, is even more important. Assessing the quality of such a design in detail is beyond the scope of this book. However, simple checks can be done, by asking for:

- Evidence that the EPC has experience in the specific industrial branch
- Integral pilot plant or commercial scale test results

Table 4.3 Checklist for purchasing pilot-planted processes

Check item	Background
The process has been pilot plant-tested	The word "pilot plant" is easily used. Find out what really has been tested or used One must perform a due diligence on Heat & Mass balances and products to confirm claims made by technology supplier
The pilot plant is an integrated down-scaled version of the commercial design	Often, only a pilot plant is available and not a commercial scale design used for the down-scaling
New equipment items	An equipment is new if it has not been used for the specific process application
Characteristic times pilot plant same as commercial scale	Has pilot plant same time scales as commercial scale, or are their time differences by different hold-up times?
Components composition in every stream	Have all pilot plant streams been analysed for component build-up?
Continuous operating time	Has the pilot plant been operated in continuous mode, and what was the longest integrated endurance run?

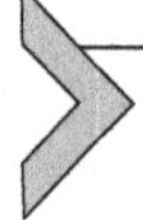

4.4 Risk assessment for mini-plant decision

4.4.1 Mini-plant choice

For processes only involving gas and liquid flows, mini-plants can be very useful. They cost only a fraction of a pilot plant and can also be quickly modified or cleaned when a problem occurs. A mini-plant typically operates at 0.1 kg/h production rate. Reactions and separations such as distillation can easily be carried out and problems such as foaming or frothing are quickly discovered, especially when the construction material is glass.

In cases where a pilot plant is skipped, a mini-plant can reduce potential risks enormously, but then the mini-plant should be a down-scaled version of the commercial scale process design, in the sense that it contains all unit operations and all recycle flows of the commercial scale design. Each unit operation cannot in all cases be a precise down-scaled version. The internals of a distillation, for instance, will be different in type from the commercial scale distillation column.

Even if a pilot plant is envisaged, a mini-plant first can be of value, as the compositions of internal flows can be determined and a flow sheet model can be validated. Unexpected side reactions, for instance in the bottom of distillation column, and azeotrope formation in a distillation can be quickly identified. Also, foaming, frothing, and solids formation will show up in such a mini-plant. These findings will be used to modify the commercial scale design and the down-scaled pilot plant design.

For pharmaceuticals and fine chemicals production, mini-plants and particularly continuous mini-plant can be very interesting. The production rate is often sufficient to produce product for clinical trials and/or market developments. The time to market is then often shortened. Scale-up later to commercial scale production can be done by numbering up the reactor pipes and/or production trains.

4.4.2 Mini-plant cost estimate

Mini-plant cost estimation is carried out in the same way as pilot plant cost estimation. A process concept design is made and equipments or pilot plant vendors are asked to provide cost estimates of the mini-plant. Some companies have in-house cost estimate methods and data for mini-plants.

4.5 Decision on integrated down-scaled pilot plant

4.5.1 Pilot plant purposes

Harmsen (2018) provides a list of pilot plant purposes. The purposes relevant for process scale-up are treated here.

The first purpose of the pilot plant for process scale-up is to discover the unknown unknowns. Any phenomenon not known or not considered will show up in the pilot plant. This can, for instance, be fouling, corrosion, erosion, catalyst decay, azeotrope formation, foaming, and particulates formation, causing all kinds of processing problems.

The second purpose of the pilot plant for process scale-up is to validate the process flow sheet model, so that each stream composition is known and each heat flow is known. The various process parameters, the known unknowns, are varied within a range established at lab and primary research, in order to find the most optimal operating window, such to obtain desired performance and ditto product quality. Sample points at various points in the pilot plant should be made to be able to validate the process model and to see whether build-up of trace components occurs.

The third purpose of the pilot plant for process scale-up is to test construction materials on corrosion. To that end, the construction material should be the same as the commercial scale design and, in addition, test construction material samples will be placed in various process streams so that corrosion rates can be determined for various process stream compositions.

The fourth purpose of a pilot plant can be to validate process control and operating procedures. This is most relevant for large-scale pilot plants containing solid (particulates) processing units, for which process control and operating procedure often are also part of the process development as standard controls and procedures are not adequate for the new process.

The fifth purpose of the pilot plant can be to provide product samples to customers to test the product performance of the new process. Even if the product is a known product for an established market and is a so-called specification product, this is an important purpose. If the process chemistry is new, or the product purification section is new, the product may contain new trace elements which may cause different behaviour for the processing of the product at the client or may cause a different behaviour of the product of client.

For performance products such as resins, polymers, paints, processed food, and pharmaceuticals (for clinical trials), this is a known purpose, but for bulk chemicals this is often neglected. In some cases, the amount of product for these tests at the client determines the pilot plant size. The alternative is to run the pilot plant for an additional time to produce sufficient number of samples. This is often costlier than increasing the pilot plant design capacity to produce sufficient product for client testing. Intimate customer relations are necessary to find out whether these product performance tests will be needed. This means that the marketing department has to be involved in the feasibility stage, so that the pilot plant capacity can be determined and its capital and operating cost. Hoyle (2002) provides additional useful information.

The sixth pilot plant purpose can be to train operating personal of the future commercial scale plant. This is particularly relevant for a pilot plant with non-standard process controls and procedures such as for pilot plants involving solids processing.

For all these purposes, it is essential that the pilot plant is a small-scale version of the commercial scale plant with the same process steps, the same unit operations, the same recycle flows, and the same conditions.

A pilot plant of a technology provider, or equipment manufacturer, has the purpose of showing the potential client that the required performance for his application is indeed obtained. Often, such a pilot plant contains only one- or two-unit operations. The scale-up method of their proprietary equipment is known, in most cases, to the technology provider. The data generated in such pilot plants is used by the equipment manufacturer to perform a reliable scale-up and to make a proper estimate for the operating costs and the investment costs, and eventually, this mitigates the risks for the investor as well as for the manufacturer. Furthermore, the scale-up data from the pilot plant can serve as a process guarantee given by the equipment manufacturer/technology supplier.

Chapter 6 describes scale-up from pilot plants to commercial scale.

4.5.2 Decision on having a pilot plant

Here are guidelines for the decision to have a pilot plant obtained from Merrow (1991). Merrow derived these guidelines having analysed a large number of scale-up projects with and without having a pilot plant.

An integrated pilot plant with all process steps and recycle flows is certainly needed if either:

A) More than four new process steps are involved.

B) The process contains one new process step and a complex recycle flow is involved.

C) The process contains a novel solids handling step.

D) The feedstock is a crude solids resource.

A new process step means not operated before at commercial scale for that particular application. The most common misunderstanding is that a process step is considered proven, because the unit operation has been applied to other applications and is wrongly considered as not new (Merrow, 1991).

A complex recycle flow is a recycle flow over at least two process units.

If this guideline is not obeyed and a pilot plant is not built, while either criterion A–D is valid, then that project very likely will end in a commercial scale disaster.

For all other project cases, an integrated pilot plant is not strictly needed, but the start-up time will be much shorter, and thus, the design production capacity will be reached much earlier, if an integrated down-scaled pilot plant is applied in the project. Time and money spent in the development on the pilot plant is easily recovered in the shorter start-up of the commercial plant.

The only sure way to avoid a pilot plant without the risks associated is to build the plant as a clone of its predecessor for the same product and the same feedstocks. This technical risk aversion, however, causes a different risk, namely that the competitors build innovative more efficient plants and they reduce the production price to get these plants fully loaded. They can drop the price to a level that they still make a profit, while the conventional plant makes a loss. So the technical risk avoidance causes a financial risk.

4.5.3 Industry project statistics of not having a pilot plant

Independent Project Analysis (IPA) has carried out extensive analysis of their large database of industrial projects. Most of those analyses have not been published, but can be obtained from IPA by contracting them. However, some of their analysis results have entered the public domain. Here are these main findings.

By analysing process projects on key technical parameters for commercial scale success regarding having a pilot plant, the author could derive following project clusters from the project results as reported by Andras Marton (2011), Merrow (1991, 1988):

Disaster projects

Projects in the IPA database that were classified as complete disaster mean that the process never produced significantly, often not even after spending double the amount of the scheduled investments. The root causes of these disasters were that the project did not have an integrated downscale pilot plant.

Merrow (1991) derived from many failed start-up projects that an integrated down-scaled pilot plant is needed if either:

- More than four new process steps
- Recycle streams over more than 1 unit
- A new solids processing step
- Crude solids feed

All cases that did not have an integrated pilot plant with one or more of the above-mentioned conditions ended in failure. It means that they only reached the design targets after spending more than 30% additional investment and several years of delay. Some were abandoned in the end.

Moderately successful projects

Projects in the IPA database that were moderately successful mean that their average start-up time is 30% more than the average over all projects, and that the standard deviation of 130% is also much larger than average of the successful projects, being 50%. These projects had also no pilot plant and had 2–4 new process steps (Merrow, 1991; Marton, 2011).

For the processes containing 2–4 new process steps, an integrated down-scaled pilot plant in most cases will be economically and in view of total time schedule a valid decision. The pilot plant cost and time will be easily made up by the shorter start-up time (and the higher production rate in the first year) of the commercial scale implementation.

Successful projects

Projects that were very successful, are projects with a start-up time of 90% or shorter than the averaged start-up time of all projects. These projects had an integrated pilot plant down-scaled from the commercial scale design and running continuously long enough to indicate reliability (Marton, 2011).

The start-up time index used by IPA is most likely the start-up time correlation by Merrow (1988), which is found in Chapter 7 of this book.

For all clarity, projects not having an integrated pilot plant, because they are not needed according to the Merrow criteria, can also be successful. This is illustrated by the cases described in Chapter 8, Sections 8.1–8.3.

Epilogue on no-pilot-plant-decisions

Bell (2005) of DuPont gives three reasons why this advice on piloting is still not always taken. The first reason is the ignorance of the publications by Merrow, proving when pilot planting pays off in time and money. The second reason is pride. Merrow presents averages and most engineers consider themselves above average; they believe they can beat the odds. The third reason is the legitimate need for innovation and a haste to get new products or new processes to the market place. The time for piloting will take months and business management may conclude that a business opportunity window will no longer be open if the process development timeline is too long. Unfortunately, to the decision makers, collapsing timeline can produce the opposite result of a very expensive start-up of the full-size plant when a better overall timing could have been achieved at far lower costs via the use of pilot plant.

4.5.4 Value of information approach for development plan decisions

Value of Information is a relatively new method for taking decisions on innovation plan options. Bos (2014) of Shell presented a version of this method developed within his company and originally used for upstream exploration decisions, but adapted for process development decisions. This method is based on a few elements. All elements are expressed in Net Present Value (NPV $) impact on the final implemented process. As the development costs have to be paid up front, this is a good way of taking these early investment elements properly into account.

Firstly, a project option is defined in which no further information is generated by a development programme, i.e. using only the information known at that point in time. Then, all identified unknowns and risks are made explicit and, subsequently for each of those risks, the consequences of a failure and an estimate of the chance of a failure are agreed upon. These are then combined into a negative expectation NPV term and subtracted from the initial NPV. I call this option: “direct implement”. Then one or more development plans are made. For each of these plans, the development costs are estimated. With the estimated information gathered, a new chance of failure is estimated. With these elements, the cost of information and the value of that information, a new final project NPV is determined and compared with the direct implement option. So, for the direct implement option and a development option, final NPV numbers are available and a decision on economic grounds to have the development or not can be taken.

The number of identified unknowns and their risks can be quite large. It is better to write them all down and work them out, rather than making a quick guess of little relevance and neglect them.

The method can also be applied when different technologies for the commercial scale process are considered, each with their advantages and risks. I have made up an example shown in Table 4.4 of a process containing an exothermic reaction section to illustrate this. Each process design option has its own (long) list of identified risks and assessment of the financial consequences of a failure due to each of these risks. Each option also has its own process development options to reduce the risks.

Table 4.4 Example of calculations value of information by development effort

	Multi-tubular fixed bed		**Fluid bed**	
Value of information element	**Direct implement**	**Development**	**Direct implement**	**Development**
Design Base NPV M$	100	100	110	110
Consequence failure NPV M$	−125	−125	−120	−120
Chance of failure %	40	40	80	80
Expectation failure NPV M$	50	50	−96	−96
Initial net NPV	50	50	14	14
Development NPV M$	0	−10	0	−30
New Design base NPV M$	100	105	120	125
New Consequence failure	−125	−125	−120	−120
New Chance of failure %	40	10	80	20
New expectation failure NPV	−50	−12.5	−96	−24
Final Net NPV	50	82.5	14	71

Project exothermic heterogeneous catalyst reaction systems.

The artificially constructed very simple example is a project with at the heart a heterogeneous catalytic reaction, which is highly exothermic. The reaction section can be designed as a multi-tubular fixed bed or a fluid bed both with cooling via cooling walls and evaporating water as cooling medium. The risk of failure is here simply lumped. In reality, all specific unknowns with their specific risks and their consequences in NPV should be identified.

The development plan for the process containing the multi-tubular reactor is to have an integrated pilot plant with a single tube. The development cost in NPV are −10 M$. The development plan for process containing the fluid bed reactor is to have an integrated pilot plant with a fluid bed reactor. In addition, a cold-flow test unit is applied to determine solids processing and to determine the heat transfer coefficient for various gas flow rates and the effect of scale-up by comparing values with the smaller scale pilot plant results. The development costs as estimated in NPV are −30 M$.

The Design base NPV for the fluid bed process is higher, because the capital investment cost is lower, and catalyst can be withdrawn and added under operating conditions, avoiding stops to replace spent catalyst. The chance of failure of the fluid bed process, without development, is however estimated to be much higher compared to the fixed bed process.

With these estimates, the values of Table 4.4 are obtained. The following conclusions can be drawn.

A: The development cost is far less than the gain in NPV, so a development effort is economically sound.

B: The final net present value of the fixed bed is higher than the fluid bed process.

As the method strongly depends on the completeness of the "risk register" and the estimation quality of each of the (many) failure consequences and the failure chances, it is of importance to have the exercise carried out by a team of experienced development experts. To avoid bias to carry out the development anyway, the team should also have experts from outside the project and also the team should be led by an experienced facilitator. The whole Value of Information exercise will take two or three team sessions with homework in between to determine the cost of plans in NPV terms. The elapsed time is a couple of months, including reporting.

The number of failure types and the number of development options are not limited in theory, but in practice should be limited to the major failure consequences and major development plan options, as each failure estimate, and chance of occurrence, is an estimate with a limited accuracy; hence, small differences in outcomes will be insignificant.

The attractiveness of this Value of Information method is that:

(a) Different process development options can be considered in a quantitative way, based on real risks and their money impact on the commercial deployment.

(b) The chances of outcome occurrence are carried out by discussions in a team by which knowledge is shared and thus the collective knowledge is increased.

(c) Different process designs, each with their R&D options, can all be compared in one go

Bos (2014) presented the method and also showed the method for a theoretical case of multi-tubular fixed bed reactor at commercial scale with following process development options:

A single tube test reactor
A multi-tube test reactor
An integrated pilot plant with a single tube reactor
A single tube test reactor followed by demonstration plant

He showed that the demonstration plant decreased the uncertainty by 95% and the residual uncertainty cost was reduced to 2 M$ for a specific risk. The single tube test reduced the uncertainty by 80% with a residual uncertainty cost of 8 M$. He did not present the cost of the demo-plant and the single tube test, but it is likely that the additional cost of the demo-plant outweighs the difference in residual risk of the single tube test. Further information on the Value of Information method for innovation is provided in the book by Artman (2009).

4.5.5 Pilot plant design

The first question to answer in the pilot plant design is: What is the purpose of the pilot plant? In general, the purpose of the plant is to validate the commercial scale process concept design and to validate the process flow sheet model. If that is its purpose, then the pilot plant should be a scaled-down version of the commercial scale plant with the same process steps, the same recycle flows, and the same conditions. Moreover, sample points at various points in the pilot plant should be made to be able to validate the process model and to see whether build-up of trace components occurs.

Often, the purpose of the pilot plant is also to test construction materials on corrosion. To that end, construction material test sections should also be built in the pilot plant so that corrosion rates can be determined for various process stream compositions.

Sometimes, the purpose of the pilot plant is (also) to provide product samples to customers to test the product performance. This is often the case for performance products like resins, polymers, paints, processed food, and pharmaceuticals (for clinical trials). Often, the amount of product for these tests is in the order of 500 kg or more. This then means that the pilot plant capacity has to be large to produce these amounts in a reasonable time. Intimate customer relations are necessary to find out whether these product performance tests will be needed. This means that the marketing department has to be involved at least at the end of the research stage, so that the pilot plant capacity can be determined and also its capital and operating cost. Hoyle (2002) provides additional useful information.

How to design the pilot plant? The most important part of the answer is that at least an experienced process developer has to be involved in the design, either an employee of the company or an employee from a company with experience in pilot plant designing and construction.

It should not be done by an engineering procurement contracting (EPC) company who only does commercial scale design and construction. I have noticed several times that such a company gives no priority to the pilot plant design and construction, because for them it is a very small project with little incentives. Such an EPC also does not know the specific requirements of a pilot plant: being of easy access to any subcomponent, easy to modify upon failure, and having many sampling points. Involving a dedicated pilot plant EPC company is, therefore, advised.

As mentioned before, the pilot plant should be a scaled-down version of the commercial scale plant, containing all process steps and recycle flows. Because of the high external surface area to volume ratio of the equipment, special precautions should be taken to insulate the pilot plant so that it has no cold walls or cold spots.

The capacity of the pilot plant will depend on the smallest feasible size of certain process unit operations and on the amount of product that has to be produced for testing by clients in a reasonably short time. For gas-liquid processes, the pilot plant capacity is often in the order of a few kg/h. For processes containing particulates, processing the pilot plant will in general be much larger as down-scaling has more limitations.

For particulates process equipment, large-scale test for the particular solids can often be carried out at the technology vendor premises. However, the integrated effects of the process as a whole cannot be tested. Also, the vendor may not be willing to perform the test with the real materials which can cause corrosion and safety issues. He may suggest surrogate materials for the testing. The usefulness of this type of testing is, according to Bell (2005), questionable.

4.5.6 Pilot plant cost estimate

Pilot plant cost estimation is similar to a commercial plant cost estimation. A cost estimate for the decision to have a pilot plant is, in general, based on the main equipment of a concept design. Cost estimates are then often obtained from pilot plant engineering and vendor firms. Large companies may have their own specialist department for pilot plant detailed design and cost estimation.

After the decision has been taken in the feasibility stage to have a pilot plant, detailed design and an accurate cost estimate will be made in the development stage.

4.6 Decision and design of cold-flow test units

For novel process equipment with complex geometries, often a so-called cold-flow test unit is constructed in which the fluid flow behaviour of the commercial scale (or near-commercial scale) can be studied. The purposes of these cold-flow plants can be:

A) To experimentally optimise the geometry to obtain the desired performance

B) To validate a computational fluid dynamics model

C) To reduce the scale-up risk by having a full-scale experimental model of the commercial process equipment

Cold-flow models are extremely useful for obtaining large-scale information at moderate cost, which otherwise cannot be obtained. This is particularly the case for particulates (solids) processing. Reliable scale-up knowledge of particulates flow behaviour is very limited and cold-flow models are often the only way to obtain reliable large-scale information about the flow behaviour.

Cold-flow models are also very useful in validating computational fluid dynamic (CFD) models. By changing geometries and flow rates, the CFD models are validated, not only for a single-point design solution, but also for their trend effects of geometry and flow rate variation. In this way, confidence for the large commercial scale process equipment can be obtained.

Critical aspects that can be studied in mock-up units are residence time distribution, mixing rates, mass transfer, heat transfer, gas and liquid hold-ups, and pressure drop.

The limitations of cold-flow models are similar to the limitations of CFD models, for gas-liquid and liquid-liquid systems where bubble and droplet break-up and coalescence play a role. The most used fluid in these cold-flow tests is water. If a gas-liquid mixture is present in the real process equipment, then often air or nitrogen is used as gas and water as liquid. If it is known that the real process mixture shows slow coalescence behaviour, then sodium sulphate is sometimes added to the water to reduce the coalescence speed of bubbles. The desired hydrodynamic behaviour, such as rapid mixing of feed streams, or plug flow behaviour can then be obtained by modifying the cold-flow geometry until the desired behaviour is obtained. For liquid-liquid systems, often water alkane cold-flow mixtures are used. Here, also the droplet formation and droplet coalescence can be very different from the real mixture. The reason is that air water and alkane water systems have bubble or droplet formation and coalescence behaviour that is different from the real mixture.

A design with all geometries and the requirements for experimentation and observations will be made. If the mock-up model has to be transparent, a dedicated construction firm should be involved. A cost estimate for the mock-up may also be obtained from this firm.

4.7 Feasibility stage scale-up pharmaceuticals, agrochemicals, and fine chemicals

4.7.1 Pharmaceuticals feasibility stage

For pharmaceutical products, the whole trajectory from research to commercial scale production is very different from other products such as for bulk chemicals. The focus is developing the new medicine product and not developing a new process. Table 4.5 summarises these major differences between pharmaceuticals and bulk chemicals.

The product, its process, and the process operation need approval by the regulatory bodies such as the Food and Drugs Approval (FDA) administration. This means that the scale-up is all about ensuring that the same product

Table 4.5 Scale-up key differences between pharmaceuticals and bulk chemicals

Aspect	Pharmaceuticals scale-up	Bulk chemical scale-up
Key economic criterion	Shortest time to market	Lowest total cost
Key success factor	Product and process quality assurance and approval	Reliable process
Process	Multi-product	Single product

is made at the large scale and that its process has been completely defined and agreed by the regulatory body.

The stages and gates involved are, therefore, also very different. From Levin (2006) and Kane (2016), following stages applied in the pharma industry are derived:

Discovery stage: The new molecule is assessed on its activity.

Pre-clinical stage: The new active molecule (product) is defined.

Clinical phase I stage: The new product and its formulation are made in a small-scale pilot plant for tests.

Clinical phase II stage: The new product is made in intermediate scale pilot plant for further clinical testing.

Clinical phase III stage: New product is made at larger scale plant for studies with many patients.

Approval stage: The product is approved by the regulatory body and decision for commercial scale production is taken.

Manufacturing stage: The new product is produced and sold.

A commercial scale process design stage up front of the manufacturing stage is not mentioned by Levin. Kane describes a case in which, in clinical phase III, the larger batch sizes could not be produced. "The wholesale process change could mean that the regulators will require the reformulated drug to start of over again at clinical phase I" (Kane, 2016). Kahn then argues that clinical phase II is the correct point at which to evaluate the scalability of manufacturing processes and the potential regulatory impact of any changes and not at the launching clinical stage III (Kane, 2016). In cases where a product's formulation needs tweaking prior to a clinical phase III trial, a bridging study can be conducted to demonstrate pharmacokinetics equivalency between the old and the new product (Kane, 2016).

The value of making process manufacturing designs in the clinical phase II stage for all subsequent stages is emerging. AstraZeneca, GlaxoSmithKline, and Foster Wheeler have founded a non-profit company Britest Ltd., whose objective is to deliver major competitive benefits to pharmaceutical and chemical companies by designing the best processes and manufacturing strategy for each member. Britest has developed methods for the development process and for determining all potential (feasible and unfeasible) process options (Ainsworth, 2005). For pharmaceuticals, the introduction of continuously operated process is not so easy as for fine chemicals. The regulatory bodies also must understand and accept the new manufacturing method. Considering batch processes only is deeply engrained in this industry. In his book on pharmaceutical scale-up, Levin in 2005 stated that scale-up is increasing the batch size. Other options than batch operation is for him beyond his horizon. It is also clear that, in that same year, Ainsworth (2005) mentions continuous processes as options to consider. So this is changing.

It is also very clear, however, from Kane (2016) and Ainsworth (2005) that commercial scale process design and also planning the steps to the final stage is best placed in the clinical phase II stage of the pharmaceutical industry. So, also here, a feasibility and planning study is placed up front of the development stage (Clinical phase III).

Scale-up methods for unit operations processes are also found in Sections 4.2 and 4.3. Additional scale-up methods for unit operations often applied in pharmaceuticals are provided by Levin (2006). Here is his list of unit operations:

- Blending and mixing of powders in tumbling blenders
- Wet Granulation
- Fluid bed spray granulation
- Compaction and tableting
- Extrusion
- Powder filling hard shell capsules
- Film coating

Chapter 6 provides information about how to perform pilot plant tests in existing pilot plants to provide reliable information for the performance of the existing commercial scale process.

4.7.2 Agrochemicals feasibility stage

The development of agrochemicals such as fungicides and pesticides is similar to pharmaceuticals. The new product also goes through a series of tests with increasing scale. The new product and its process have to follow regulation procedures. The reasoning of the pharmaceuticals section, therefore, also holds to a great extent for agrochemicals.

4.7.3 Fine chemicals feasibility stage: Unit operations choices in view of scale-up

The choice of process unit operations is a very important part of the feasibility study of fine chemicals. Often, a short time to market is an overriding economic criterion. Traditionally used are mechanically stirred vessels, with cooling by solvent evaporation with a reflux and operated in batch-wise. Often, the scale-up is carried out using the empirical scale-up method and using a series of pilot plant reactors increasing in scale. Often, the recipe changes to counteract the observed negative scale-up effect. This empirical scale-up method then still takes some time.

If an easy to scale-up reactor type is chosen and such a cooled pipe reactor operated continuously, then the scale-up time can be shortened. BASF; Kleiner (2011), Lonza; Schnider (2018), and GSK; Gonzales (2001) have applied such cooled pipe reactors for fine chemicals production. Kleiner (2011) presented a case for a polymer development that using a continuously operated cooled microflow reactor, instead of a mechanically stirred tank reactor operated batch-wise, had a net present value of presented 100 M€ over a development in which mechanically stirred tank reactor operated batch-wise was to be applied. The large reduction in innovation cost and time was mainly due to not needing an intermediate demonstration scale, as the product quality uncertainty was removed by the reliable microflow reactor.

Schwalbe (2002) showed for a fine chemicals case that the investment of a micro-reactor of 0.4 M€ had a payback time of 0.2 years and had savings of 2.3 M€/year over the existing mechanically stirred batch reactor, due to higher product yield on feedstock and 50% less labour cost. Christian Schnider (2018) of Lonza Switzerland presented experimental results of a fluorination in a continuous pipe reactor with a volume of 15 L, with large safety and cost advantages over a mechanically stirred batch reactor of $4\,m^3$.

So, continuously operated cooled pipe reactors, introduced early in the concept stage, can reduce the time to market. The examples show that, for fine chemicals, this is already happening.

Microflow reactors are also advocated for the pharmaceutical and fine chemical industries. The advantages are:
- They can be applied at laboratory scale, because of their size and the small quantities of feed needed.
- Scale-up is easy as heat transfer limitations are absent, so scale-up can be done by numbering up the number of channels with the same dimensions and flow rate of the lab-scale reactor.

Information on designing and constructing micro-reactors and micro-plants is provided by Fink and Hamp (2000) and Wiles and Watts (2016). However, although micro-reactors are available now for decades, commercial scale applications are hardly reported. Even Fink and Hamp (2000) and Wiles and Watts (2016) do not report commercial scale applications. Perhaps, concerns about cleanability and risk of fouling play a strong role in the decision not to go for this technology.

Hulshof (2013) has gathered 240 industrial cases of fine chemicals scale-up. The book is out of print. Anyone who can get hold of a copy should do so. Probably, he can draw learnings from these cases for his own projects.

4.8 Planning for concurrent development and commercial scale design

When the business case is clear and the incentives for commercial scale production are large, there is often pressure from top management to shorten the time to market. In general, this can cause cutting corners, bypass stages, and stage-gates and in the end, according to Merrows (2011), cause a commercial scale implementation failure. However, there is a safe way of reducing the time to market. This is by concurrently making a Front-End-Engineering-Design (FEED) and even the first part of detailed process engineering at the EPC contractor in parallel to the pilot plant testing, by assuming that the pilot plant test will validate the process design and the process design model without surprises. If that indeed happens, then time is gained. If, however, the pilot plant test shows flaws in the process design, needing commercial scale redesign (and additional pilot plant testing), then the FEED and detailed engineering must be redone. Then no time is gained, and extra costs have occurred. Risk of those extra costs can be estimated up front of the decision of concurrent development and design and evaluated against the extra gain if the shorter time to market is successful.

4.9 Feasibility stage-gate evaluation

The feasibility stage-gate evaluation is very critical as there a decision is taken to spend a large sum of money on the development stage. To reveal the importance of this stage-gate to all involved, Table 4.6 can be of help. By filling in for each risk item, the risk chance, and the risk effect by the stage-gate panel, the risks of omissions become clear. The undesired risk effects can then be remedied by actions as part of the feasibility stage, so that the second stage-gate can be successfully passed.

4.10 Pitfalls feasibility stage

Pitfalls for the feasibility stage, according to Merrows (2011), are:

- Greedy lead sponsor causes poor development
- Project schedule is under too high pressure
- Too little spent on Front End

Greedy lead sponsor causes poor development. In the feasibility stage, the business case and the economic incentive of the project become clear.

Table 4.6 Risk assessment template feasibility stage

Risk item	Chance	Risk effect
No safety assessment		
No health assessment		
No environmental assessment		
No economic assessment		
No technical feasibility assessment		
New process chemistry not identified		
Product is new, due to new chemistry, or process not identified as new		
No social acceptance assessment		
No front-end-loading design		
Process model validation not defined		
Scale-up methods and validation not defined		
Pilot plant not down-scaled version		
Business case not defined and agreed		
Stakeholders aspirations not understood		
No strategic fit		

The greedy lead sponsor may object to the development cost needed for development. He may force the project development team to cut too much on planned cost. This then will cause later a poor detailed design and a commercial scale implementation failure.

Project schedule is under too high pressure. The project schedule defined in the feasibility stage for pilot planting and EPC will have a span of several years. Due to a large incentive, the project timing may be forced to be shortened by reducing or even omitting pilot plant test runs.

Too little spent on front end. The feasibility study with its commercial scale design may reveal knowledge gaps that need to be closed by additional research. The decision to perform this research with its additional cost may be neglected. This can have negative consequences for the final detailed design, even if the pilot plant tests are successful. For instance, making an elaborate process model, allowing optimisation in the feasibility stage, commercial scale design may not be possible, due to the lack of parameter value information. This, in turn, then means a suboptimal design, with a large loss of economic potential for the commercial scale design and operation.

Pitfalls pharmaceuticals

Ainsworth (2005) provides following pitfalls:

- Not designing for the commercial scale at clinical phase II stage
- Not evaluating continuous process options at clinical phase II stage

CHAPTER FIVE

Development stage

5.1 Mini-plant engineering, procurement, construction, and testing

When, in the feasibility stage, a decision has been taken to have a mini-plant, then engineering (detailed design) will be executed, followed by procurement (purchasing the equipment) and constructing the mini-plant. After that, the mini-plant will be operated and tested for its performance.

The mini-plant should be a downscaled version of the commercial scale design containing all process units and all recycle streams. In addition, it may contain construction material samples in various parts of the process to obtain corrosion information.

The following design documents will be needed to start the engineering, derived from Vogel (2005) and summarised by Harmsen (2018):

- A process flow sheet with mass flow compositions
- An engineering flow diagram
- All equipment types and key element sizes
- Objectives of instrumentation
- Specifications for insulation of all pipes and equipment
- List of sample points and measuring points
- Safety concept and logic (how sensors are to be linked to actuators) and safety procedure

The actual engineering, construction, and procurement (EPC) may be carried out in-house or by an EPC contractor. For the latter, it is recommended to involve a company specialised in mini-plant and pilot plant EPC. If an EPC contractor for large-scale projects is involved, then often the project will be delayed, as it will get a lower priority than a large-scale project. Also, the quality may then be poor.

The testing plan will depend on the purposes of the mini-plant. If the mini-plant replaces a pilot plant, then it is even more important that all purposes of the mini-plant are used to define the mini-plant test programme.

Industrial Process Scale-up
https://doi.org/10.1016/B978-0-444-64210-3.00005-6

5.2 Pilot plant engineering, procurement, construction, and testing

The engineering (detailed design) of the pilot plant can start from the same design documents as provided for the mini-plant in Section 5.1. Most design documents will have been produced in the feasibility stage to determine the investment cost estimate for the pilot plant, see Section 4.5.4.

If those design documents are not available, then these must be generated, starting from a basic design to be made as a downscaled version of the commercial scale design. If that design is not available, then first a commercial scale design has to be made, as described in Chapter 4.

The engineering, procurement, and construction (EPC) may be carried out by an external contractor. In that case, it is advocated to select an EPC contractor dedicated to pilot plants only. If an EPC contractor is chosen, who also does commercial scale processes, then it is likely that the pilot plant project will get a lower priority compared to commercial scale projects. Also, the contractor will have less experience with aspects critical to pilot plants, such as easy access to sampling points and easy to make modifications.

The testing programme needs to be planned. If the pilot plant contains, for instance, a catalysed reactor, then long-term testing is needed to be able to observe catalyst decay. If the pilot plant involves food processing, then also long-term testing may be needed for the following:

- Protein/product fouling of filters, resins, membranes
- Spoilage (microbiological decay) in time
- Ageing of membrane/resins

If the pilot plant contains recycle streams, then also the long-term testing is needed to be able to observe any build-up of tracer components. In making the planning, the start-up time correlation of Merrow found in Chapter 7 can be used to get an estimate of time needed to reach steady-state operation. It may sound strange that the start-up correlation for commercial scale is also useful for the much smaller pilot plant, as in a pilot plant changes needed to solve problems can be more quickly made. But a considerable part of solving problems in a pilot plant are the same as in the commercial scale plant, namely finding the cause of the problem and making a modified design to solve the problem.

Pilot plant operators also generate knowledge by their experiences. Most of this knowledge will be tacit and cannot be coded. However, by involving pilot plant operators in the commercial scale process start-up, this tacit knowledge is likely to be helpful in preventing problems and in quickly solving start-up problems.

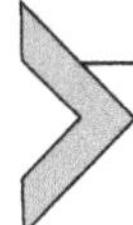

5.3 Cold-Flow unit engineering, procurement construction, and testing

The cold-Flow unit EPC will in most cases be carried out by a dedicated external firm. The test instrumentation, such as gamma ray measurements and fast cameras, will be obtained from specialist firms. There will be an extensive calibration programme of the actual measurements, to make sure that the measurements are meaningful, so that fluid flow models can be validated experimentally.

5.4 Front-end engineering design

At the end of the development stage, a front-end engineering design (FEED) is made. Bakker et al. (2014) shows that the project value is created at the early stages up to, and including, the development stage, and that in the next stages (EPC and implementation), this value must be maintained. The FEED is a very important element in preserving the created value, as it is the base for the detailed design. A poor FEED means in the end a poorly performing commercial scale process. Key deliverables for the FEED derived from Bakker and translated for process design are:

- Design Engineering Package
- Investment cost estimate (+/− 10%)
- Updated hazard and operability (HAZOP) description
- Updated risk assessment using all development results and market survey results

The objective of the Planning of the EPC is to be able to control the project by allocating resources, such as human, material, and financial, and giving time constraints to each task. For large complex development projects, this planning is the base for cost and timing estimates. Bakker et al. (2014) provides more detail on this subject.

For large projects, first, a FED-2 cost estimate with an accuracy of 20% often is made to decide whether to continue or not with the project. If the project is then stopped, a few million may have been saved. The additional advantage can also be that the FED-2 is presented to top management and that then they are involved earlier than when a FED-3 is made. They may, for instance, decide that they do not want to take the risk of the investment and ask for a much smaller capacity plant with a lower investment. It is recommended to make a FED-2 cost estimate for the smaller capacity and report the result including the return on investment for the smaller and the larger capacity to top management again. They, then, may reconsider their decision to go for a smaller capacity.

5.5 Development stage-gate reporting and evaluation

Bakker provides the following list of deliverables for the Front-End-Development (FED-3):

- Evaluation reports of the development stage (mini-plant, pilot plant, mock-up results)
- Design Engineering Package
- Investment Cost estimate (+/− 10%)
- Updated Hazard and operability (HAZOP) description
- Updated Risk assessment using all development results
- Planning of engineering, procurement, and construction effort and schedule
- Process start-up plan

In addition to this list, aspects concerning safety, health, environment, economics, technical feasibility, and sustainability (SHEETS) must be evaluated and reported.

For safety evaluation, the Dow Fire and Explosion Index as reported by AIChE 1994a is suitable.

For health evaluation, the Chemical Exposure Index as reported by the AIChE, 1994b is suitable.

For the environmental evaluation, a Life Cycle Assessment is most appropriate. ISO 14044 method is the most reliable published way of performing this evaluation and is publicly available, see ref. ISO, 2017.

The economic evaluation should include the market development, the expected sales price, and production capacity planned.

The technical feasibility should be based on a risk assessment. It is best carried out by a group of experienced process engineers and chemists. Table 3.1, of Section 3.1, can be used as a checklist for items to cover.

The sustainable development contribution has also to be addressed, as most companies have a strategy to contribute to the sustainable development goals (SDG) for 2030, as defined by the United Nations in cooperation with the World Business Council for Sustainable Development; UN, 2015.

For each of the main dimensions, social, economic, and environmental, at least one goal must be contributed to. The SDG is listed here:

Sustainable development goals:

- Goal 1. End poverty in all its forms everywhere
- Goal 2. End hunger and achieve food security and improved nutrition and promote sustainable agriculture
- Goal 3. Ensure healthy lives and promote well-being for all at all ages
- Goal 4. Ensure inclusive and equitable quality education and promote lifelong learning opportunities for all
- Goal 5. Achieve gender equality and empower all women and girls
- Goal 6. Ensure availability and sustainable management of water and sanitation for all
- Goal 7. Ensure access to affordable, reliable, sustainable, and modern energy for all
- Goal 8. Promote sustained, inclusive, and sustainable economic growth, full and productive employment, and decent work for all
- Goal 9. Build resilient infrastructure, promote inclusive and sustainable industrialisation, and foster innovation
- Goal 10. Reduce inequality within and among countries
- Goal 11. Make cities and human settlements inclusive, safe, resilient, and sustainable
- Goal 12. Ensure sustainable consumption and production patterns
- Goal 13. Take urgent action to combat climate change and its impacts*
- Goal 14. Conserve and sustainably use the oceans, seas, and marine resources for sustainable development
- Goal 15. Protect, restore, and promote sustainable use of terrestrial ecosystems, sustainably manage forests, combat desertification, and halt and reverse land degradation and halt biodiversity loss
- Goal 16. Promote peaceful and inclusive societies for sustainable development, provide access to justice for all, and build effective, accountable, and inclusive institutions at all levels
- Goal 17. Strengthen the means of implementation and revitalise the global partnership for sustainable development

5.6 Pitfalls development stage

Only the novel unit operation of the complex process is piloted and not the whole process.

The pilot plant has large intermediate catalyst hold-up vessels, not present in the commercial scale process design. This hold-up vessel containing homogeneous catalyst masks catalyst decay.

The pilot plant is run for a limited time for which slow build-up of trace elements, catalyst decay, fouling (e.g. membranes), and/or ageing effects are not observed.

CHAPTER SIX

Development stage scale-up from existing pilot plants

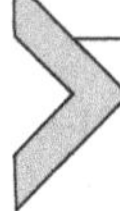

6.1 Scale-up from existing pilot plant to existing commercial scale process

6.1.1 Scale-up continuous process from existing pilot plant to existing process

Companies that have both existing pilot plants and existing commercial processes also will have a practice to do the scale-up for a novel product to be produced or a novel feedstock for the existing product. I can only point at pitfalls for these cases.

The most common pitfall for these cases is not to recognise that the process is new, and thus, not to take the steps needed to implement the "new" process; so, not to perform a potential risk assessment highlighting all the differences related to the production of the new product in the existing process and taking precaution measures to reduce the identified risks from the risk assessment.

6.1.2 Scale-up for batch process from pilot plant to existing commercial scale process

Companies with these facilities will have their own practices to do the scale-up to the commercial scale. There are, however, two pitfalls for these companies.

The first pitfall is the same as described in Section 6.1.1 of not recognising that the process is new.

If the process is recognised as new, then the guidelines on down-scaled in paragraphs below may be of use.

The second pitfall is not to consider evaluating new continuous processes with far lower production cost and higher reliability. Evaluating alternative process options including continuous processes is treated in Section 4.7.3, where a fine chemicals case is shown of having an existing commercial scale

Industrial Process Scale-up
https://doi.org/10.1016/B978-0-444-64210-3.00006-8

process and still the additional investment of a new process is economically more advantageous than the existing process, because of much lower feedstock and operation cost.

Down-scaled temperature-time profile batch reactor

The product yield and by-product formation are strong functions of the temperature-time profile. Keep, therefore, the temperature-time profile in the pilot plant the same as in the commercial scale plant. There are two options available:

1. Making a dynamic model of the commercial scale plant with the reaction enthalpy and kinetics as input.
2. Simulate experimentally the commercial scale plant by having the same specific surface area for heat transfer and the heat transfer coefficient and keeping the heating and cooling medium temperature the same as the commercial scale plant.

Ad 1 A dynamic model based on reaction enthalpy estimate (from similar reactions), the maximum heat transfer capability expressed as the heat transfer capacity per unit reaction liquid volume ($W/(m^3s)$) and the feed rate of the limiting reactant, can be used to calculate the temperature-time profile of the commercial plant. This profile should then be mimicked in the pilot plant. Optimisation of product yield can be obtained in the pilot plant by varying the limiting reacted feed rate (below the maximum set by the commercial scale model estimate).

Down-scaled crystallisation in existing pilot plant from existing commercial plant

For performing a down-scaled crystallisation in the pilot plant, not only the temperature-time profile should be the same as the commercial scale, but also the supersaturation concentration profile and the local supersaturation at the cooling surface (if cooling by heat exchange area is applied).

The volume averaged supersaturation can be kept the same by having the same turbulent eddies for micromixing and by having the circulation time for macromixing the same. The first can be achieved by the same energy dissipation per unit fluid volume in the pilot plant and in the commercial scale plant. The circulation time can be kept the same on both scales by dividing the fluid volume by the pumping flow rate of the stirrer.

It should also be noted that the residence time-scale and occurrence of multiple crystal-forms and the desired crystal size distribution for individual crystallisation processes are major selection criteria for the type and geometry of the crystalliser and its internals for the commercial scale and its down-scaled pilot plant.

6.2 Scale-up from existing pilot plant to new process

Often, a pilot plant is available from a technology provider or equipment manufacturer, who can do pilot-tests for his clients for novel products and/or novel feedstock. The pilot plant can be a single piece of equipment or a unit operation. The technology provider may have many different unit operations, which can be connected so that an integrated complex process is formed at pilot scale level. The question is: How to reliably scale-up from such an existing pilot plant to the commercial scale process to be designed?

Scale-up starts of course at the laboratory level, where the physical properties, limiting/degradation temperatures, physical appearances, (vapour-liquid) equilibrium data, and physical behaviour are investigated. The dedicated lab-studies give an indication on the suitability of a certain unit operation for a specific application and the type of pump required. Furthermore, the lab-studies give an initial operating window to be investigated in the dedicated pilot-unit operations. The reader is referred to Chapters 3 and 4 for more details on basic design knowledge generation in the discovery and concept stage.

The most suitable scale-up method from existing pilot plants, firstly, starts with the critical scale-up phenomena described in Section 4.2.2. For the pilot plant at hand, an analysis is then made which of these scale-up phenomena are relevant for the specific application envisaged. Secondly, for the identified phenomena, a suitable scale-up method is selected from Section 4.2.3. Thirdly, that method is then validated for the specific application by pilot plant tests. Fourthly, the commercial scale design is then made by selecting the same equipment type and by applying the selected scale-up method available with the technology provider or equipment manufacturer.

Biomass extrusion scale-up example

If, for example, the pilot plant is a reactive extruder in which ligno-cellulosic biomass is heat-treated to an intermediate product suitable for enzymatic hydrolysis of the cellulose, the critical scale-up phenomena are then:

The scale-up method of choice for the pilot plant will be hybrid: model-empirical-based, as the biomass is not a uniform plastic material

Table 6.1 Example scale-up phenomena, method, and validation tests biomass pre-treatment in extruder

Relevant scale-up phenomenon	Scale-up method	Validation test
Residence time distribution	Hybrid Model-empirical	Tracer injection response
Shear rate distribution	Hybrid Model-empirical	Torque measurements
Shear to heat transformation	Hybrid Model-empirical	Temperature profile measurement
Heat transfer	Hybrid Model-empirical	Energy balance heating and extrusion

over the radius of the extruder and not over the length of the extruder. Model-based scale-up with a validation on a single size pilot plant would make scale-up behaviour prediction to a very large extruder very uncertain. Model validation over two scales would make scale-up reliable, if the model is based on fluid flow basic principles and not just is a simple correlation of the performance versus scale using two measurement points of the two scales.

The model should be validated for each scale-up phenomenon by specific measurements as indicated in Table 6.1. In addition, it is advised to produce product samples for different temperature, residence time profiles, and different screw speed. These tests will reveal the sensitivity of the product quality on these parameters and allow for optimisation.

The commercial scale extruder will be of the same extrusion screw shape. The diameter and the length of the extruder will be much larger. Using the model for the residence time distribution, the residence time distribution will be kept the same or even closer to plug flow. Using the model, the temperature profile versus residence time in the extruder of the biomass will be kept the same as the pilot plant, as this is very critical to the product quality. Model calculations will be needed to arrive at the same temperature profile. To that end, the total energy dissipation by extrusion per unit biomass feed will be kept the same, as well as the skin temperature for heating. For adjustments and optimisation, the screw speed of the commercial extruder will be variable.

CHAPTER SEVEN

Implementation stage

7.1 Introduction to implementation stage

This chapter treats all critical elements of the implementation stage. The implementation stage involves a decision by the manufacturing company on the commercial scale capacity and location, i.e. whether the first-of-its-kind process will be demonstration scale or a full scale. Then an engineering, procurement, and construction (EPC) contractor has to be selected. The detailed engineering, procurement, construction, and commissioning will be carried out by the EPC contractor in cooperation with the manufacturing company. The start-up plan, organisation, execution, and post-start-up reporting, will be carried out by the manufacturing firm. The chapter ends with a list of pitfalls of the implementation stage. This stage is also called execute stage in Oil&Gas companies.

7.2 Rationale demonstration scale decision

A demonstration scale is the first-of-its-kind small-scale commercial process. The purpose is to reduce the technical and business risks of the new process. In particular, when the market for the product is under development and medium-paced market growth is foreseen, a decision for demonstration scale process first may be more quickly and easily taken. The Shell GTL Bintulu plant described in Section 7.4 is an example of such a decision.

There can be another reason for having a demonstration scale process. It is likely that a first-of-its-kind process design will have some redundancy introduced by the process designer to make sure that the new process will work and reach its design capacity. As a senior process engineer remarked to me: "rather lead than dead". One can do this by oversizing equipment or even adding equipment that might be needed. In one case, I know, even an extra distillation column, with a bypass option, was installed and never used.

Industrial Process Scale-up
https://doi.org/10.1016/B978-0-444-64210-3.00007-X

In the demonstration plant, this redundancy is less expensive than in the full scale. Moreover, by operating the demonstration process and performing test runs, this redundancy can be determined and can then be avoided in the full-scale design. This redundancy can easily be 10%–20% of the investment. It, of course, strongly depends on the culture of the in-house process engineers making the FEED.

Experience curves, also called learning curves, of processes of the same industry branch can also be used to make crude estimates of the cost reductions of subsequent plants. Business risks with a new process are sometimes uncertainty of the market size (increase) of the product, or strategic risks; the company is not sure that it will pursue aggressive sales increase, required to get the full-scale process loaded. Sometimes top management is uncertain whether the research and development has covered all technical scale-up risks and decides to have a demonstration scale process first.

The demonstration scale should be so large that the process still makes a profit, although the capital charge per ton of product, due to the small scale, is higher than that for the full scale. If this is not the case, then top management may lose interest in the whole new process and will not be willing to invest in the commercial scale process later. Typically, its capacity is 10% of the full-scale process and the capital expenditure is typically 30% of the full scale. So if it fails, partly or totally, maximally only 30% of the capital investment is lost.

Here is a story from my own experience. Top management, when seeing the investment of the commercial scale of a first-of-its-kind process for the first time, decided that they did not want to carry the risk for such a large sum and then asked for a smaller capacity process with a smaller investment. When that design was made, and they then saw that the Return on Investment was much lower, they asked for an improved design for the same capacity that had the desired return on investment. This then forced the development team to redesign the process, which required additional testing.

This delay in development time and extra cost for redesign and piloting can be avoided by making demonstration scale design and a full-size design in the feasibility stage and presenting both the investment cost and return on investment of both scales to top management so that, early in the innovation funnel, a decision can be taken on whether a demonstration process is desired or not.

Demonstration scale processes are typical for first-of-its-kind oil and gas and metal processes, where the full scale requires an investment in the order of 10 billion €. A demonstration scale, a factor 10 smaller in investment, is then often decided upon. An example is the Shell GTL process described in Section 7.4.1.3.

In bulk chemicals, demonstration scale processes are rare. For gas-liquid processes, they are virtually absent. BP chemicals, for instance, has developed and implemented in the last 20 years four new processes; to note: the CATIVA acetic acid (Jones, 2000), the leap vinyl anhydride monomer (VAM) (BP, 2002a), the AVADA process (BP, 2002a), the linear alfa olefins (LAO) process (BP, 2002b), and the PTA process (BP, 2015). All commercial scale processes were directly scaled-up from a pilot plant to full capacity with scale-up factors exceeding 10,000 and investments ranging from 100 to 250 million $. Drawing from my experience, I conclude that in bulk chemicals, demonstration scales rarely occur. If they occur, it is for a new polymer for which the product market has to be developed, such as the CARILON case described in Chapter 8.

By monitoring the demonstration scale process performance in detail, a large database can be made. Test runs and debottlenecking can increase the database further and allow for cost reductions later for the commercial scale design.

For a first-of-its-kind commercial scale process, demonstration scale or full scale, the same steps are followed as described in the next sections.

7.3 Engineering design, procurement, and construction

7.3.1 Contractor selection

According to Merrow (2011) page 254, "that decision for contractor should absolutely be made by the project director and the team. Any involvement made by the sponsor's purchasing organisation in the selection of contractors is likely to be a catastrophe." The background to this guideline is that the project director knows which EPC firms have experience in the specific branch and a good track record. If the sponsor is from a different branch with experience of a particular EPC contractor for his branch, then the decision for a wrong decision by the sponsor is easily made.

The importance of choosing an EPC contractor with experience in the branch and also choosing subcontractors for certain equipment with experience in the specific branch of the process cannot be overemphasised. For instance, an EPC contractor with experience in bulk chemicals is not suited for food processing as he will not know the critical details for success such as no dead ends (not even very small) where microorganisms can grow and contaminate the product. Vice versa is also true. An EPC subcontractor with experience in the food industry is not suitable for the bulk chemicals industry. He, for instance, may not know that the bulk process must run un-interrupted and reliably for 6 years. He may propose a technology that needs regular replacement of an element, such as the filter cloth in crystallisation process case of Section 8.2.

The overall quality of the final detailed process design, the purchasing of equipment, and the construction and erection of the plant determine, of course, the final reliability of the process. A description of detailed measures to ensure this is beyond the scope of this book. Involving an engineering contractor experienced in the specific process is a key factor for success. If the process is very novel, then an engineering and procurement contractor experienced in processes involving similar products and streams should be searched for. If that is not possible, then at least an engineering contractor experienced in processes of the industrial branch should be searched for.

7.3.2 Detailed design

Detailed design will in most cases be carried out by the EPC contractor in cooperation with the manufacturing company. For large projects, a well-structured organisation will be set up. Bakker et al. (2014) provides a complete description. Here, only the organisational structure will be summarised. The manufacturing company will form an asset development team with a venture manager, an operations manager, and a technology manager, to which a project manager from the EPC contractor will be added. The Steering committee from the manufacturer will steer this team. The project manager will then form a team with experts in finance, quality, engineering, contracts, planning, manufacturing, and procurement.

Bakker spends a large chapter on project risk management. The reader is referred to that chapter for all the details. Here are some important findings and critical success factors of EPC work:

- Even low-uncertainty projects usually come with failure and delays
- Previous studies have shown that high-risk projects are not less successful than low-risks project
- Important success factors are: trust, market, teambuilding, leadership
- Risk identification must be done by the right people
- Risk monitoring and taking appropriate measures are needed to achieve project goals.

7.3.3 Procurement, construction, and commissioning

Procurement is about purchasing equipment and services. This is also a critical step for the start-up success. Contracting with explicit liabilities and guarantees are essential elements, but even more important is the proven record of the purchased equipment, as the cost of a failed process can never be paid by the vendor. Here again comes the importance of having an EPC contractor who knows which subcontractors deliver reliable equipment and services for the particular process branch. Bakker provides lots of advice on contracting and the advantages of certain types of contracting over others (Bakker et al., 2014).

The construction of the process is an enormous project in itself. For the Shell GTL Pearl process, for instance, with an investment of 20–21 billion $, 52,000 workers were involved (van Helvoort et al., 2014). Quality of planning, scheduling, monitoring, and management is of utmost importance. Providing guidelines for this part is outside the scope of this book. Bakker provides guidelines for this step (Bakker et al., 2014). Most knowledge is, however, found inside the EPC contractor organisations.

Commissioning the process, i.e. cleaning all process equipment and testing all instrumentation, piping, and vessels, is also a very critical step. Here, errors in the construction will be found and be corrected. This commissioning is, in general, carried out in close cooperation between the EPC contractor and the manufacturing operational staff and the operational staff is in the lead.

A pre-start-up audit marks the end of the commissioning step. It is mandatory in the oil & gas industries to prepare a statement of fitness (SoF). It is a handover document and signed by the project manager and the future process owner to confirm that all process safety requirements have been addressed (Bakker et al., 2014). More details on project close out are also provided by Bakker et al. (2014).

7.4 Start-up commercial processes

7.4.1 Start-up time prediction

7.4.1.1 Start-up time Merrow industry correlation

A very important critical factor for a successful start-up is having an integrated down-scaled pilot plant. Merrow of independent project analysis derived the following guideline for the need of such pilot plant by analysing a large number of scale-up projects with and without having a pilot plant.

An integrated pilot plant with all process steps and recycle flows is certainly needed if either:

A) More than four new process steps are involved

B) The process contains one new process step and a complex recycle flow is involved

C) The process contains a novel solids handling step

D) The feedstock is a crude solids resource

A new process step means not operated before at commercial scale for that application. The most common misunderstanding is that a process step is considered proven, because the unit operation has been applied to other applications and is wrongly considered as not new. A complex recycle flow is a recycle flow over at least two process units.

If this guideline is not obeyed and pilot plant is not built, then either criterion A–D is valid that the project will end in a commercial scale disaster. This means that the commercial scale plant will have a much long start-up time than the industry average. It may even never produce the required product in significant quantities.

The start-up time correlation of Merrow (1988) is:

$$t_{start-up} = 3.3 + 3.7\,N - 3.2\,F + S$$

$t_{start\text{-}up}$ = the start-up time in months, starting when real feedstock is fed and ending when steady state is obtained.

N = Number of new process steps. A step is new if it has not been used for that specific application.

F = Fraction of mass and heat stream compositions that are known. F ranges from 0 to 1.

S = Solids processing factor.

S = 0, if only gases and liquids are processed.

S = 0.7, if refined solid products, such as plastics, are produced from gas and liquid feeds.

S = 10.8, if raw solids, such as mined ores, are fed and processed.

The start-up time starts when for the first time all feed stocks are fed to the process. So the pre-commissioning time with static and dynamic testing of equipment and control is excluded from this start-up time.

The number of new process steps, N, means the number of process steps that have not been applied at commercial scale for that particular process. So even if a process step has been used elsewhere for a particular application and is now for the first time used for a different substance or different stream, then that process step is new. A new process step is defined here as process section, like reactor section or purification section, containing new technology.

The factor F stands for the fraction of materials and heat streams that are known. It means the fraction of process streams the size and composition of which are truly known. In general, this knowledge is obtained using process simulation programmes that have been validated experimentally by integrated pilot plant's long duration test runs, where stream samples have been analysed on composition and compared with simulation predictions. In particular, recycle streams with slow build-up of trace components should have been analysed and materials of construction have been tested for these streams.

The correlation has been obtained by Independent Project Analysis from 40 process start-up cases, including liquid-gas processes, refined solids process plants, and raw solids process plants. The $R^2=0.93$ and the standard deviation for the start-up time of the correlation = 2.4 months.

Merrow remarks that the complexity of the process appeared not to be a significant parameter, as soon as the number of new process steps was used as parameter.

It should be stressed that this correlation is based on all kinds of process innovation project start-ups, regardless of the quality of the R&D, Design, and start-up preparation. In the next section, the effect of these other factors on the start-up time is shown.

7.4.1.2 Critical success factors reducing start-up time

In my quest for finding methods to reduce the start-up time, I found in the literature 10 critical success factors for start-up. These are shown in Table 7.1. I then defined a dimensionless start-up time, the actual start up time being divided by the Merrow correlation start-up time of Section 7.4.1.1. I plotted the, thus, obtained dimensionless start-up time of 10 industrial cases of the company I worked at in 1992 and a graph similar to Fig. 7.1 emerged. The original graph was lost in the mist of time. I reconstructed it from my memory. The values for the first three critical success factors have a large uncertainty. Merrow stated that if a pilot plant is needed according to his analysis, but is not applied, then the start-up may result in a process that never produces product in significant quantities though the start-up time is infinite. Most data of the 10 cases belong to the higher Critical success factor numbers, so there the dimensionless start-up time values are more certain.

Table 7.1 Critical success factors for commercial scale process start-up and their numbering

Critical success factors [5]	Numbering
For development and design	
Process is defined as new?	1
Integrated down-scaled pilot plant available?	2
R&D and EPC knowledge integration?	3
Scale-up knowledge for unit operations available?	4
For start-up preparation	
Potential problem analysis carried out?	5
Precaution measures taken?	6
Complete start-up team?	7
Operators trained for start-up and operation?	8
Start-up plan?	9
Documentation	10

Harmsen (2018, p. 70) and Harmsen (1996d).

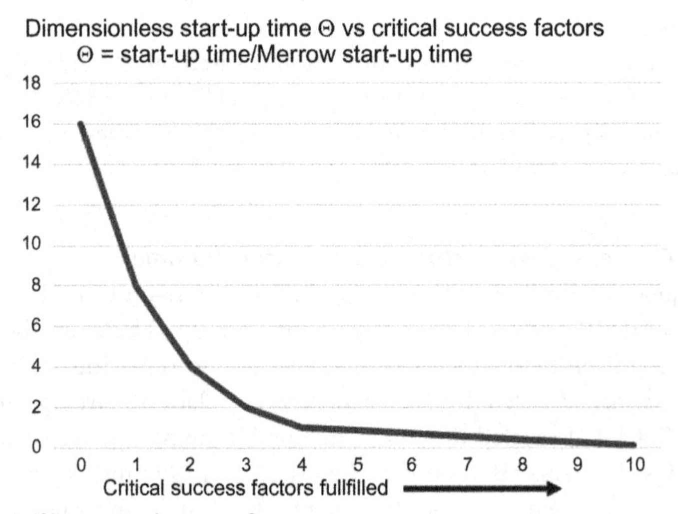

Fig. 7.1 Fulfilling critical success factors.

No critical success factors taken into account

If none of the critical success factors are fulfilled, then the start-up time will be very long or even infinite. This happens when the process is not even recognised as new. This can, for instance, happen when conventional unit operations are applied and Merrow's definition of new is not applied to that specific process. It can also happen if a new catalyst is applied in an existing process and everybody involved thinks that nothing is new and no precaution measures are taken. I have noticed this latter reasoning twice in my career, resulting in very long start-up times and large production losses.

Process is defined as new

If the process is recognised as new, then it is likely that a development is started and knowledge is generated reducing the start-up time. If, however, according to Merrow, an integrated pilot plant is needed and not applied, then start-up time will still be very long or even infinite.

To stress this point of recognising newness, the definition of new is stated here again: A process (step) is new if it has not been in commercial scale operation for that feed, that product, that catalyst, that piece of equipment, or those conditions.

Integrated pilot plant applied

If an integrated pilot plant is needed and applied, then the start-up time will be considerably reduced. It may fall within the uncertainty range of the Merrow correlation. Merrow, in a further analysis of start-up results, focused on what had been done on process development and, in particular, whether an integrated pilot plant was applied. He found out that an integrated pilot plant with all process steps and recycle flows is certainly needed if either:

A) More than four new process steps are involved

B) The process contains one new process step and a complex recycle flow is involved

C) The process contains a novel solids handling step

D) The feedstock is a crude solids resource

A new process step means not operated before at commercial scale for that particular application. The most common misunderstanding is that a process step is considered proven (not new), because the unit operation has been applied to other applications and is wrongly considered as not new. A complex recycle flow is a recycle flow over at least two process units.

In cases where this guideline was not obeyed, and pilot plant was not built, that project often ended in a commercial scale disaster. This meant that the commercial scale plant had a much longer start-up time than the industry average, with significant extra cost to get it working. In several cases, it never produced product in any significant amounts of product (Merrow, 1991). Marton of IPA, in addition, reported in 2011 that if no pilot plant data were available, then the average actual production rate was 20%–70% of the design production rate in the second half of the first year after the initiation of the start-up, while with pilot plant data available the production rate was 50%–100% in that same time slot (Marton, 2011). So, with a pilot plant the production rate after start-up increases a lot compared to no pilot plant data available.

R&D and EPC integrated

Merrow (1991) and Marton (2011) both report that integration of the R&D results into the EPC design and construction results in reduced start-up time and increased productivity in the first year of commercial scale production. So, this is also a critical success factor to be fulfilled. In Fig. 7.1**7.1**, the dimensionless start-up time is estimated to be 2. Marton indicates that his dimensionless start-up time ranges from 0.6 to 2.5 if pilot plant data are used in the EPC, so the dimensionless start-up of 2 is a reasonable estimate for fulfilling this critical success factor (in addition to the first two).

Scale-up knowledge unit operations available

Vogel (1974) of Dow Chemical reported that scale-up knowledge of unit operations is essential for successful implementation. In combination with the three previously stated critical success factors, the dimensionless start-up time of the few cases in my company ranged then around the value 1.

Critical success factors for start-up preparation

Critical success factors for start-up were obtained from Ryan (1972), Fulks (1982), and Hendersen (1985). They were sequenced in the order of execution time and then the dimensionless start-up time was plotted vs these critical success factors. I just counted the number of these critical success factors and plotted the dimensionless start-up time vs that number and obtained a downward straight line. When the total number of critical success factors was 10, the dimensionless start-up time was 0.1. So, the right-hand part of Fig. 7.1 is obtained by a combination of data and logic.

Potential problem analysis carried out

Fulks (1982) and Hendersen (1985) both stress the importance of carrying out a potential problem analysis to prepare the actual start-up. The reason for this additional potential problem analysis is that, often in the long EPC time, the project context may have changed. Also, now the operating people are involved.

Such a potential problem analysis is best carried out by a start-up support team, involving at least an experienced process operator, a process engineer, a process developer, and a research chemist.

If the process is moderately new and a similar process is already in operation, then a potential problem analysis is even more important, as then the natural habitude is to take the start-up as easy. In my career, I noticed in one start-up case that the process was considered conventional and no potential problem analysis had been carried out and not even a start-up team had been formed. The start-up turned out to be very problematic. As this was not expected and no start-up support team was in place, the problems could not be quickly analysed, so that the start-up time increased even more.

If the process is moderately new, then a potential problem analysis can be carried out by highlighting the differences with the existing process. Differences should not only be looked for at the process technology, but also at the context of the process; what is different in the location of the new process compared to the existing process. The list of 15 modal aspects of reality presented by Harmsen (2018) on page 95 can be of help in this respect. Relevant in most start-up cases are differences in social acceptance; language and communication habits; physical differences; weather and climate; kinematics such as geometry and plot area, and numerical; and the process capacity and size.

Precaution measures taken

The potential problem analysis should be used to define precaution measures for the start-up procedure. Hendersen (1985) states the importance of precaution measures. He mentions ensuring that the product can be delivered in time, for instance, by having product from an existing process in store. An alternative precaution measure is to promise the product to be delivered at a much later time than the actual start-up.

Prior to the start-up is a pre-commissioning and a commissioning step in which all equipments are cleaned and controls are tested. Tests are often carried out with water, solvent, and nitrogen. This step can also be a precaution measure. Often, flaws in the construction are detected in this step.

Complete start-up team and organisation

Firstly, the organisation of the pre-commissioning, commissioning, and start-up is treated. Lager (2012) discusses four start-up organisation models:

(1) Final production organisation does also the plant pre-commissioning, commissioning, and start-up.

(2) Project team does the pre-commissioning and hands over the process to the operator for start-up.

(3) Project team also does start-up and afterwards hands over the process to the operator.

(4) Final production organisation forms an intermediate integrated organisation with the project organisation, who does the project, the pre-commissioning, and start-up. Afterwards, the intermediate organisation is dismantled and the operational organisation remains in place.

After examining several start-ups in which Lager was personally involved, or people of his network were involved, he concludes that for simple small process scale start-ups the first model works best. For large-scale start-up, the fourth organisation works best.

From my experiences, I conclude that the first organisational model can be also very successful for large complex processes, but only if an advisory start-up team is connected to the operational organisation at least a year in advance of the start-up date. The advisory start-up team must contain at least the process engineer and the researcher if the process is novel. It may also contain a process control specialist and an analytical chemist.

Start-up leader

A very important factor of the start-up organisation is the start-up leader. According to Lager (2012), an experienced start-up leader determines 90% of the success. I agree with this observation. The start-up leader should have experience in start-ups, preferably as start-up leader. During start-up, major events can occur at any time, requiring action from him or her. To have a round-the-clock leadership, she or he should have a start-up leader assistant, who can replace him or her during the time off. Also, in this way, the company trains future on-the-job start-up leaders.

Start-up support team

The start-up support team, also called the flying squad, should be ready to quickly analyse occurring problems and help to solve it (Lager, 2012). It typically consists of an experienced process operator, a process engineer, a process control specialist, a process developer, and a chemist involved in the R&D of the project.

Operators trained for start-up and operation

Fulks (1982) and Hendersen (1985) both emphasise that the operators should be trained in the actual process and in the start-up procedure. Nowadays, dynamic process models are available with the same interface as the actual distributed control system, by which process operators can be trained. Training can also be obtained by using the pilot plant. Training sessions in which development process engineers explain the basics of the process will also be of help. If the process is a modification of an existing process, then operators of the existing process can help to provide training or, even better, the operators of the new process obtain training at the existing process.

Chapter 8, Section 8.3, describes the case of not sufficiently trained process operators with a start-up time of over 3 months, while the process had no new elements and all mass and energy flows were known. So the dimensionless start-up time was 30.

Start-up plan

The start-up procedure should be defined in detail, be written down, and distributed to all involved and also comments to procedure should be invited from all involved in the start-up, so that in the end a commonly carried procedure, or start-up manual, is in place.

Documentation

All documentation of the process should be complete and available to all concerned. Also, the start-up plan should be available long before the actual start-up, so that it is part of the start-up training. It also ensures that all have the same and complete information. It avoids, thus, misunderstanding and miscommunication.

During the start-up, important events and changes needed to overcome problems also must be documented to avoid misunderstanding and miscommunication.

After the start-up is completed, a post-start-up report should be prepared for several reasons.

First of all, if a second process project is started, then all errors and learning points of the present new process can directly be found and consequently the second design can be improved of the first implemented design.

Secondly, several of these post-start-up reports can be analysed for common errors and learning points and can then be used to improve the innovation quality procedure in the company.

The post-start-up report should be written very soon after the start of the start-up; preferably, when the start-up team is still present so that start-up team members can contribute to the report. The report should at least contain the following points:

- The actual start-up time and interruptions before steady-state operation is reached
- All deviations from the design performance in terms of product quality, product production rate, and utility requirements
- All deviations from design specifications and conditions
- An analysis of the relations between the performance and the deviations

7.4.1.3 Start-up time and critical success factors Shell GTL case

To illustrate the combined power of the Merrow start-up correlation and the critical success factors, The Shell GTL scale-up case is taken. The historic development of GTL technology in Shell is well-described by van Helvoort et al. (2014). The process contains 10 sections:

- Air Separation Unit (ASU) to produce pure oxygen
- Shell Gasification Process (SGP), thermal partial oxidation of natural gas to syngas
- Hydrogenation Manufacturing Unit (HMU) additional hydrogen production
- Heavy Paraffin Synthesis (HPS) catalytic conversion of syngas to hydrocarbons
- Heavy Paraffin Conversion (HPC) catalytic hydrogenation of heavy paraffins to lighter linear paraffins
- Synthetic Crude Distiller (SCD) to distil into naphtha, kerosene, and gasoil fractions
- ReDistiller Unit (RDU) distilling to waxy raffinate
- HydroGenation Unit (HGU)
- Detergent Feedstock Unit (DFU) Separating hydrogenated hydrocarbons into LDF and HDF fractions
- Wax Production Unit (WPU) Separating hydrogenated hydrocarbons into Wax fractions

In the process development, only the HPS was tested in a pilot plant. The SGP, derived from oil gasification to syngas, was tested in a large commercial scale burner at Hoechst Celanese in Houston (p. 246). These tests, however, were not fully representative of the Bintulu conditions, as Celanese includes a carbon dioxide recycle to the feed of the SGP. That this posed a risk of short burner life times in Bintulu plant was anticipated in that additional measures could not be ruled out beforehand.

Additional tests were also carried out on the removal of trace components from the syngas at Chemische Nijverheid Ostende, Belgium. This resulted in additional guard beds in the syngas section to remove traces and to deal with process upsets.

The Bintulu plant with a capacity of 14,700 bbl/day (550 kton/year) started up in 1993. Table 7.2 shows that the actual start-up time was more than a factor 2 longer than the Merrow start-up prediction.

Table 7.3 shows the analysis with the critical success factors for the GTL Bintulu plant. None of the critical success factors for development and start-up were obeyed. This explains the very long start-up time.

Table 7.2 Shell GTL Bintulu start-up time

Parameter	Value	Comment
Number of new steps; N	10	Helvoort [4] p. 243 shows 10 sections
Mass and heat flows known; F	0.5	Gas composition of Shell Gasification Process is not completely known
Integrated pilot plant needed?	Yes	10 New steps with complex recycle
Integrated pilot plant available?	No	Only synthesis reactor section pilot plant
Merrow start-up time; months	39	Industry average correlation for $N=10$ and $F=0.5$
Actual start-up time; months	96	Start-up 1993; no steady state till 1997. After explosion Restart 2000 and steady state 2004; p. 252

Table 7.3 Shell GTL Bintulu Helvooort [4] assessment with Critical Success factors [5] p.70

Critical success factor [5]	Project assessment Shell GTL Bintulu van Helvoort et al. (2014)
For development and design	
Project is defined as new?	No. Only Synthesis section is considered new. SGP feed is new due to different feed composition, but not seen as new pp. 245–246. It caused very short burner life times p. 252
Integrated pilot plant available?	No. Only SMDS section was pilot-planted p. 191
EPC contractor experienced in process branch?	No. The chosen EPC, JGC Corporation (p. 244), is an upstream Oil and Gas contractor
R&D and EPC knowledge integration?	No. Only on synthesis section. pp. 255–256
Scale-up knowledge unit operations available?	Not for Shell Gasification Process p. 238
For Start-up preparation	
Potential problem analysis	No, Helvoort [4] p. 256
Precaution measures taken	No, Helvoort [4] p. 256
Complete start-up team	Not reported by Helvoort [4]
Operators trained for start-up and operation	Not reported by Helvoort [4]
Documented start-up plan	Not reported by Helvoort [4]
Overall documented info and decisions	Not reported by Helvoort [4]

Table 7.4 Summary start-up problems, root causes, and corrective actions

Problem	Root cause	Corrective action
Stress cracking corrosion CO/CO/ organic acids streams	No long-term pilot plant testing	Post-weld treatment and replacement carbon steel to stainless steel
Metal dusting syngas train	No long-term pilot plant testing	Adapting conditions and metal passivation
Wax solidification underground pipes	Heat tracing failed	Above ground carbon steel pipelines
SGP burners very short life time	No long-term pilot plant testing	New burner design
Operating highly heat-integrated process with many shutdowns	No integrated pilot plant tests	Learning by doing
Coked up hot-oil system	Poor design	Replace complete hot-oil inventory
Safeguarding causing shutdowns	Individual safeguarding	Integral safeguarding redesign methodology
Equipment failure	Equipment not commercially proven for this application	Strong interaction with vendors to solve problems
Many complete shutdowns	Single train little redundancy	Extra attention to upkeep
Utility from outside not available	No potential problem analysis prior to design	Proactive study 1 year after start-up. Additional boiler capacity installed
Explosion air separation unit	Carbon dust burning causing alumina oxidation, causing explosive liquid oxygen vaporisation	Ultra-filtration air, liquid oxygen pumps, bath-type vaporiser, improved monitoring

Shell GTL Bintulu plant 14,700 bbl/day (550 kton/year) from van Helvoort et al. (2014).

Table 7.4 provides a summary of the problems, the root causes, and the remedies of this first-of-a-kind GTL process, as reported by van Helvoort et al. (2014).

For the subsequent Shell GTL Pearl plant with a factor 10 larger capacity, the commissioning and start-up took less than 7 months (van Helvoort et al., 2014, p. 292). The actual start-up time is not provided. The 7 months include the commissioning step for each process section, which probably is the major part of the 7 months. In the design and start-up of that plant, all learning points from the Bintulu plant were taken into account.

The EPC contractor was now a JGC/KBR joint venture Helvoort [4] p. 284. The Merrow start-up time for the Pearl case with $N=0$ and $F=1$ yields a start-up time of less than a month.

This case shows that, having no new process steps and by following the critical success factors, the actual start-up of even a megaproject can be very short. It also confirms the point of Merrow that process capacity and complexity are not relevant parameters for start-up time.

7.5 Pitfalls implementation stage

Pitfalls for implementation have been obtained from Bakker et al. (2014). He presents the following causes of project failures:

- Unrealistic plan
- Unrealistic budget
- Lack of understanding stakeholders' aspirations

Similar pitfalls but in different wording are provided by Merrow's key mistakes of megaprojects (Merrow, 2011) for this stage:

- Project schedule is under too high pressure
- Fire project managers overrunning schedule
- Business-deal-only driven
- Shaving 20% of EPC budget
- Contractors should carry the risk

As the implementation stage is by far the highest investment part of the whole innovation project, the temptation to cut cost in this stage is understandable. It requires strong feet to resist this temptation and keep pointing at the importance of having a reliable process for the production. It may help to point at the fact that 50% of process innovation projects of the oil & gas sector fail and other sectors have similar failure rates (Bakker et al., 2014). All these failures can be prevented by good project management, using available guidelines as presented in this book.

From my own experiences on start-ups, I have observed the following pitfall.

The company organisation thinks that the process to be started up is a conventional process, already in operation at other locations of the company. They then think that no specific attention needs to be paid to the start-up preparation other than instructing the operators for normal operation. Because of the lack of preparation with no potential problem analysis and no robust start-up plan, the differences and novelties of the process are not identified, and the start-up goes horribly wrong.

CHAPTER EIGHT

Industrial scale-up cases

8.1 Liquid-liquid extractive reaction by Taguchi model-based product quality control

While working as an advising technologist at a manufacturing plant of my company, I became involved in a very interesting scale-up case. The existing process was a batch process in which an organic chloride component was converted with an aqueous alkaline solution into a hydrocarbon and a salt. The organic component was in an organic liquid phase and the alkaline was in the aqueous phase. The alkaline was transferred to the organic phase, reacted, and the salt transferred back to the aqueous phase.

Because of market growth for the product, more production capacity was needed. To save operator cost and capital cost, a continuous process was considered and developed. In the new process, a solvent was used and a new homogeneous catalyst to speed up the reaction. The chemistry and reaction were studied experimentally in a batch laboratory scale reactor of about 1 L.

A novel continuous reactor, a multi-stage mechanically mixed column with many chambers and each chamber with a stirrer mounted on a central vertical axis, was chosen, because my company had experience with this type of equipment for liquid-liquid extraction for refinery applications and design rules were available. The multi-stages were needed to keep the required reactor volume low as the required conversion was over 99%. Fig. 8.1 shows a sketch of the process.

The following mass transfer and reactions take place in this process. Alkaline is transferred from the aqueous phase to the solvent phase. There it reacts with the organic chloride to a hydrocarbon and a chlorine salt. The latter is transferred back to the water phase.

All design stages with more and more precise economic evaluations were carried out and all looked very attractive and the go ahead for detailed design and construction was given to an engineering contractor. A scale-up risk was considered to be the precise RTD of this multi-stage mixer. Therefore, a

Industrial Process Scale-up
https://doi.org/10.1016/B978-0-444-64210-3.00008-1

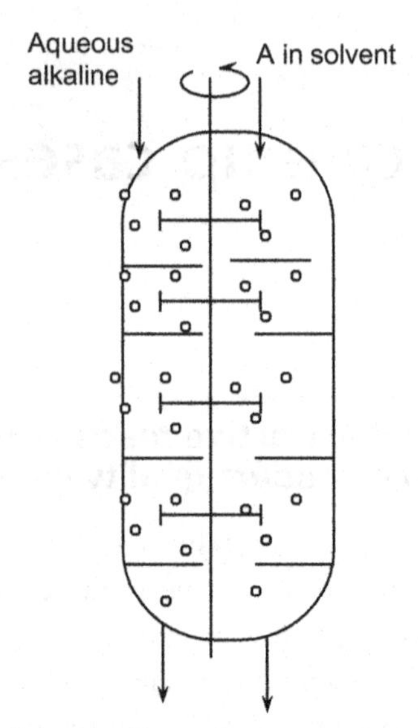

Fig. 8.1 Liquid-liquid reactive extraction process.

cold-flow model was bought from the same supplier as the commercial scale design. The RTD was tested using salt pulse injections and measuring the conductivity response curve at the inlet and the outlet. It appeared that the resulting RTD curve was very similar to a theoretical curve with the same number of well-mixed vessels in series as the actual stages in between the horizontal baffles.

Just to be sure that the clients would have confidence in the product from this new process, a large batch of product was made using the new recipe and sent to the client. The product batch was well within the product specifications. The organic chlorine content, for instance, was 20 ppm, while the product specification was 180 ppm.

The client, however, reported that he was totally dissatisfied with the product from the new process because, in the years he had received the product, the average value of the organic chlorine content was 165 ppm and the standard deviation was less than 5 ppm. The product he wanted to have was a product which on average had a value of 165 ppm and with a standard deviation equal or less than the values of products delivered so far.

The process technologist came to me and said I cannot fulfil this requirement of the client because the continuous reactor will be fed from batch reactors; the inlet organic chlorine concentration varies by 10% at least so that the output concentration will also vary a lot. But, even worse,

I cannot steer at the wanted average value of 165 ppm, because it takes 4 h to determine the organic chlorine content and, every 3 h, a batch reactor will start to feed to the continuous reactor. What to do?

I proposed to use the Taguchi method in which robustness to input disturbances is obtained by design and not by control. I proposed the following: *We* will make a complete steady-state model of the reactor using the kinetics and mass transfer behaviour, which predicts the outlet concentration as a function of the remaining design parameters, stirrer speed, alkaline concentration, and aqueous alkaline flowrate and temperature with their upper and lower limits. We will treat the average organic chlorine feed and its standard deviation as a given. We will try to find a set of design parameters values such that the input variation has minimal effect on the outlet concentration variation. This we call the Taguchi method.

The Taguchi concept of robust design is shown in Fig. 8.2.

The derivatives needed were calculated with the model. The model was first validated with experiments in a mechanically stirred batch reactor. It appeared that both mass transfer and kinetics were playing a role and that the model coefficients could be determined.

Because the model is highly non-linear, the conditions for robustness were easily found, and we were lucky that the standard deviation of the output concentration could indeed be reduced to about 5 ppm at the average value of 165 ppm. The robust set of parameter values appeared to be (Harmsen, 1994, 1996b, c): maximum stirrer speed, maximum alkaline concentration, and aqueous flow rate such that alkaline feed was 10% higher than stoichiometric feed. The temperature was then chosen such that the target value of the organic chlorine content in the product would be obtained. A statistical process control quality control method was developed

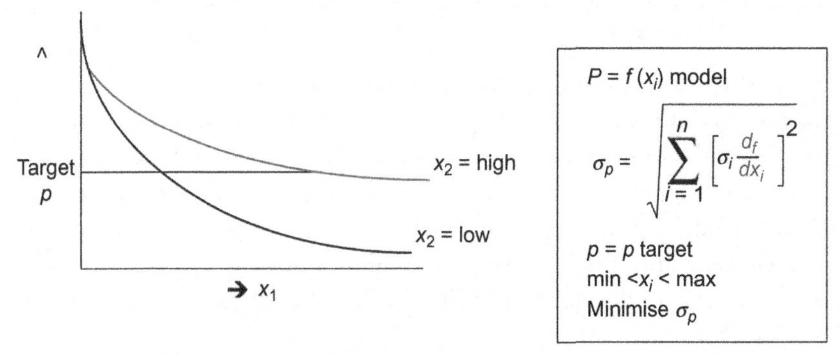

Fig. 8.2 Taguchi robust design to variation in x_1 for design parameters x_1 and x_2.

for the temperature to steer any systematic long drift away from the average organic chlorine content value back to the desired value.

The whole method was explained to the operators prior to start-up. Just before the start-up, the RTD of the commercial scale reactor was measured and showed that, indeed, theoretical number of mixed vessels matched the number of chambers. This gave confidence that the engineering construction was well executed.

The start-up was very successful, and the client was very satisfied with the product quality delivered.

A general description of using models to obtain Taguchi robustness is provided (Harmsen, 1996a).

Learning points

LP1: Product specifications are only a small part of communications with clients.

LP2: Scale-up of a single unit operation without pilot plant can be successful, if all knowledge essential for the performance is available and integrated in a validated model.

LP3: Taguchi robustness by design method, combined with a process model, is a powerful tool to create output robustness to input variation.

Ad LP1: The client is a very important stakeholder in process innovation, as discussed in Chapter 2.

Ad LP2: This learning point is addressed in Chapter 4.

Ad LP3: The power of using design models is addressed several times in the book.

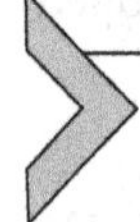

8.2 Bulk chemical product start-up conventional process

As advising technologist, I was asked to prepare and guide the start-up of a large crystallisation process to produce 100 kt/a. The process consisted of a large crystallisation section of two crystallisers in series, each with external heat exchange loops and a rotating drum filter with spray washing to remove the mother liquor. The mother liquor was treated in an isomerisation reactor to convert the unwanted isomer to the desired isomer. The reactor outlet was sent to a second smaller crystallisation and filtration section to recover the desired product.

The process was a copy of an existing process plant in a different part of the world, but of the same company. I visited that process and asked the operating personnel about their start-up procedure and how long it took to reach the product quality specification. They said: "Start-up is easy, we start-up under the same conditions as normal operation and within 24 h the product meets its market specification of 99.9995% purity". I said: "So you have 300 t of off-spec products, what do you do with that". They said: "We dissolve that in the second plant into the liquid (to be recrystallised). We have a special provision for that". Then I realised we had a problem, because we did not have this re-dissolving option. I asked why it did take a day to reach product specification, but nobody knew the answer.

I went home and made a very simple dynamic model of the crystallisation section together with a process control expert. The model consisted of a kinetic rate expression for nucleation with an assumed exponent of super saturation of 10 and a crystal growth kinetic expression linear to the super saturation and linear to the total surface area. The model had only two particle sizes, very small nucleation particles and averaged crystal particles. The kinetic rate constant of the nucleation kinetics was obtained by fitting the model to the steady-state result of the existing plant. The crystal growth rate constant was obtained from assuming that mass transfer was the limiting step and that the Sherwood number was 2.

We ran the model at the standard conditions and we saw that formation of very fine crystal particles occurred in the second vessel first. It took around 24 h to reach the final steady-state normal particle size. We considered this fine particle formation to be the cause of the product quality problem, due to poor washing at the rotating drum filter. The reason for this fine particle formation was also easy to understand. During the start-up, product was fed to the crystallisation vessels. Because the second vessel operated at a lower temperature, super saturation first occurred there. Because no crystals were present, the super saturation grew fast causing a very high nucleation rate and consequently small particle formation. We then started the model run again, but now at equal temperatures in both vessels. Now, nucleation started first in the first vessel, but the fine particles dissolved in the second vessel. After some time, super saturation also occurred in the second vessel, but because of the particles fed from the first vessel only particle growth occurred, and steady-state particle size was rapidly obtained. After a few hours, we slowly adjusted the temperatures in both vessels until the final temperature conditions were obtained. We further optimised the start-up conditions to rapidly obtain the desired particle size without fine particle feeding to the filtration section. The model was further extended with the control loops and was then used to train the operators. At the start-up of the new process, product specification was met within 1 h.

By the way, the off-spec amount of that 1 h of production, some 10 t, was used for making an epoxy resin not requiring the high product purity. So in the end no waste was produced.

Learning points

LP4: The so-called carbon copy of the process was not a carbon copy because the local context with the re-dissolving option of the reference case was different from the new process.
LP5: The dynamic model appeared to be extremely useful for optimisation of the start-up procedure and for instruction of the operators.
Ad LP4: Assuming that the process is conventional, while in reality it is new, is a major cause of start-up problems in general. By visiting the reference case plant, the newness was obtained. Chapter 4 contains details to determine whether a process is new. Chapter 7 describes the value of a potential problem analysis, focussing on differences of the design compared to the existing process.

8.3 Start-up conventional process in South Korea

As advising technologist and with process start-up experience, I was asked to help to start-up a bulk chemicals process in South Korea. The process was of a conventional design and very similar to a process already in operation in the Netherlands. The process is, however, very complex with three very different reaction sections: many separation sections, with liquid-liquid extraction, crystallisation, filtration, many distillations and stripping and absorption sections, and a prilling tower to cool and solidify the final product into prilled particles. The process has, furthermore, many recycles and one of the feedstocks, also used as solvent throughout the process, solidifies at 40°C, so that the whole process is insulated and has electric trace heating of every pipe.

Many of the process operators had no experience with process operation. Only the shift foremen and the process manager had process experience.

The start-up took more than 3 months before steady-state operation was obtained. There were many causes for this slow start-up. The first reason was that several technical pieces of equipment failed. Because of the complex process with thousands of pieces of equipment, this in most start-ups was occurring. Most equipment failures are quickly discovered in the commissioning stage and then it takes a few weeks to replace the failed equipment with good equipment. So this was not the main reason.

The second reason was the inexperienced operators. Some operators made very strange mistakes because they had very little knowledge of processes. For instance, during a night shift, a small distillation column in which feedstock was recovered from heavy end by-product was filled with liquid also containing the heavy ends. This dirty stream was then fed back to the reactors causing contamination of many process streams and off-specification production.

What may have had an additional negative effect is that the Korean management wanted to have large numbers of operators for each shift to be able to rapidly correct problems. They created these large numbers by having each shift working for 12 h and in between each shift only 8 h rest. This caused exhausted process operators, who were less alert to noticing operation deviations. This exhaustion was enhanced by the long start-up time.

The third reason was the combination effect of the first and second reason. Here is an example of this combined effect. Some glass lining of pipes was poorly manufactured, and operators were not careful with the glass-lined piping and stood on it or dropped tools on it. This caused rapid corrosion and leakage. The glass-lined pipes had then to be replaced by new pipes, by which the pipe insulation had to be removed and reinstalled. Some insulation was, however, damaged and not properly repaired by the operators; probably, because they did not know the potential consequence of this poor insulation. Upon heavy rainfall, water leaked through the damaged insulation and cooled the pipes below 400°C, by which the pipe content solidified and blocked the flow. Then the whole process had to be stopped, the pipes had to be opened at the flanges, and by hot water injection the feedstock was removed.

Learning point

LP6: Process start-up with operators with no or little process knowledge is asking for long start-ups and many problems. This learning point is incorporated in Chapter 7.

8.4 Purchasing novel process

A company purchased a novel process from a different company. The latter company had done all necessary research and development and had a demonstration plant in operation of 15 kt/a.

The research included chemical kinetics determination, a CFD model of the reactors. The development included a fully integrated pilot plant with all recycles. Extensive long-duration test runs had been performed. During the long-duration runs, the homogeneous catalyst remained active and no fresh catalyst supply was needed. Commercial scale design and start-up for a capacity of over 400 kt/a should be reliable.

A first contract was signed, and my colleague Rene Bos and I were asked to check the scale-up reliability. We discovered that the pilot plant and the demonstration plant reactors were different from the commercial scale design. The commercial scale plant reactor was a very large-scale cross-flow bubble flow reactor in which vertical baffles were placed to obtain a narrow RTD. The pilot plant and the demonstration plant reactors consisted of compartments connected via pipes with bends. The required conversion of feedstock in the reactors was 99.99995%, as the specification limit for the feedstock in the final product was 0.0001% and separation of feedstock from product did not occur in the separation section.

I postulated that the commercial scale reactors could show a very small shortcutting flow, causing a less deep conversion and consequently off-specification production. It took a lot of effort to convince the engineers that due to the very high target conversion, even a tiny fraction of shortcutting can cause the outlet concentration to increase by one order of magnitude. We asked the providing company to run their CFD model using the particle tracking option and look for shortcutting flow, even if it was less than 0.0001%. This was done and indeed the shortcutting showed-up. The baffle configuration was changed, and the pipe outlet was now taken into consideration (Harmsen and Rots, 2009). Then the CFD model showed that the shortcutting was no longer occurring.

The pilot plant design was then inspected in detail. It appeared to contain a very large hold-up vessel in the catalyst recycle loop. This vessel was so large that an enormous surplus of catalyst was present and the stated conclusion by the provider that the catalyst had not deteriorated, because no fresh catalyst addition had been needed, appeared to be wrong. In fact, the catalyst had degraded a lot. Inspection of certain effluent and recycle streams appeared to contain components resulting from catalyst degradation. A more stable catalyst was then developed.

Learning points

LP7: When purchasing process technologies, it must be checked on all details and, in particular, on those with scale-up effects.
LP8: Be aware of relatively large hold-ups in pilot plants which could mask problems.
LP9: Inspect all pilot plant streams on trace components build-up.
Ad LP7–LP9: Table 4.4 contains a checklist for purchasing novel process equipment.

8.5 Polymerisation process start-up

In my company, an old suspension polymerisation process consisting of 36 batch reactors was replaced by a new process of 6 batch reactors. The new process was a carbon copy of a process already in operation, in my company in a different country. The process was started up and was then run at 70% of its capacity for 2 months as planned by management until the regular maintenance stopped. After the maintenance stopped, the process was attempted to run at 100% of its design capacity, but this appeared to be impossible. Moreover, several times, the product was off-specification because it contained too many blue-coloured particles. After 3 months, an external expert team was formed, of which I was a member, to analyse the problem and propose solutions.

After 1 month, most problems and their causes were identified. The main problem was not reaching design capacity: It appeared that the design team knew that it was highly unlikely that the plant would be able to run at the design capacity, because for that capacity seven reactors would be needed. But with seven reactors, the capital expenditure would be so high that the required return on investment would not be met and the whole project would not have been agreed by management and, therefore, the design team had decided to choose six reactors. The design team, however, had allocated an empty space in the process plot for the seventh reactor, with the officially stated argument that was done for future capacity increases.

The problem of the blue particles appeared to be caused by having removed in the maintenance stop the revolving paddles in the top vapour part of the reactors, because the plant engineers thought that the paddles had no function. I explained that this revolving paddle acts as a droplet catcher and coalescer, so that droplets do not reach the top condenser of the reactor and the condenser stays clean. Without the coalesce, the droplets reach the condenser. Some particles stay there. In between batches, the reactors are cleaned with a blue spray and so these particles also obtain a blue colour. In subsequent batches, some particles drop back into the suspension, causing the off-specification product.

The main reason that the plant manager and engineers had not been able to identify the causes of the problems was that they had no experience with starting up a new process. They considered the whole start-up as part of normal operation. They had not made a start-up plan. There was no start-up team. The advising process technologist was not involved at all in the start-up preparation and the actual start-up. This, in turn, caused the problems to be not clearly written down and analysed. So they were not prepared at all.

Learning points

LP10: Have the process design audited and challenged by experienced people outside the design team, even if the process design is a carbon copy of an existing plant.
LP11: Appoint a manager experienced in new process start-ups as start-up leader.
LP12: Form a start-up team including engineers, process control, and process chemist.
LP13: Make a start-up plan and share it with the operating people.
LP14: Record changes and the actual process behaviour in comparison with design conditions.
Ad LP10–14: Start-up preparation and execution guidelines are found in Chapter 7.

8.6 Wastewater novel design and implementation

For the styrene monomer propylene oxide (SMPO) process, Shell wanted to explore novel process options using the expert system software package PROSYN and the PDC services. A team of Shell process developers and PDC experts on process synthesis and PROSYN was formed and first the scope of the process synthesis problem was defined. A lengthy discussion was held, whether the wastewater treatment should be part of the scope or not. The experienced Shell SMPO process developer saw no benefit in including the wastewater treatment, because this problem had been addressed many times in Shell and finally solved by choosing crystallisation from five technology options and this was implemented commercially. I argued that perhaps the mechanism of including the wastewater treatment synergy with other process elements could be discovered. PDC argued that they had special modules in PROSYN to tackle the wastewater problem, so in the end the wastewater treatment was included in the scope.

In the process synthesis, firstly, various modules of PROSYN were used. This revealed that distillation could be an option. The senior process developer of SMPO was asked to provide a list of key components representing the wastewater composition, suitable for a complex distillation analysis. He provided a list of 23 components. When using the distillation calculation module with these components, only a complex distillation concept set-up with two columns appeared in which a side draw of the first column at a tray had to be fed to a particular tray of the second column and also a side draw from that column had to be fed to a particular lower tray to the first column. Only then clean water and brine containing the organic and salt components could be produced.

A complete flow sheet simulation was carried out to size the column and determine the capital expenditure. It appeared that the distillation set-up required a factor 3 less capital expenditure and similar primary energy amounts compared to the crystallisation.

The distillation set-up was quickly tested in a small-scale glass laboratory distillation set-up. Two problems were encountered. The distillation showed foaming. This was counteracted by adding an anti-foaming agent. The operation appeared to need pH control to keep the phenol component in the salt form. When these two problems were solved, the research and development manager went to the project leader, involved in the detailed design and construction of a SMPO plant, considered to be a carbon copy of an existing plant, and explained to him that he could chose to have a distillation for the wastewater treatment instead of the carbon copy of the crystallisation section. He also highlighted that the crystallisation of the existing SMPO plant had various operational problems, while he expected that the distillation operation was more reliable. The project leader agreed to have this project plan change. He then involved a subcontractor specialised in designing dirty water evaporation processes. The concept design of the two columns was translated into four columns with multi-effect evaporation and heat integration and additional measures were added, such as a water jet at locations where fouling could occur. A patent was also obtained (de Bie et al., 2001).

The distillation section was started up prior to the main process start-up, using wastewater of the existing SMPO plant at the same location. It took less than 3 days to have the section in steady-state operation and meeting all design requirements.

Learning points

LP15: For concept designs make the scope as wide as possible, even if the process to be designed is considered a carbon copy.
LP16: Concept design using the expert system PROSYN in combination with experienced process developers can lead to novel concepts.
LP17: For a single unit operation, a pilot plant is not needed if a well-designed laboratory set-up is used.
LP18: A technology provider and subcontractor with experience in a certain field are of great value.
Ad LP15: It may take some effort to convince managers that wastewater treatment and other treatment parts also, where conventional solutions are available, should be included in the new concept design. But the effort is justified.
Ad LP17: When and when not to have a pilot plant is described in Chapter 4. This learning point agrees with that description.
Ad LP18: This learning point supports the advice on the use of external parties in research and development provided in Chapters 3 and 4.

8.7 Carilon engineering polymer new product, process, market, and strategy change

Shell discovered in 1982 the chemistry to make a novel polymer named Carilon. It is a polyketone engineering thermoplastic made from ethylene and carbon monoxide using a special homogeneous Palladium catalyst. After an enormous research and marketing exploration effort, a small production plant was started up in the Carrington, UK, in 1996. Shell then build a 55-million-pound Carilon production plant in Geismar, USA, which started up in 1998. The product was meant for application in the appliance, automotive, and electrical industries.

In 1999, Shell Chemicals announced change in strategy by selling 40% of its business, including Carilon. Shell was, however, unable to sell the Carilon business and announced in February 2000 that it discontinued the Carilon production (McCoy, 2008). In 2002, SRI announced that Shell had donated the Carilon patents to SRI International, a leading non-profit research institute (ICIS, 2002; Omnexus, 2002).

This is a clear example of what can go wrong with a new product development, in combination with a new market, and a new process, in combination with changes in business strategy.

Learning point

LP19: Developing a new product in combination with a new process and new market outside the core company business is a very risky thing to do.

8.8 Purchasing a commercially proven rotating filter

Here is an example from my own experience on purchasing a commercially proven technology. We started up a new bulk chemicals process in which the only novel element was a rotating filter with a filter cloth. For all other parts, the process was a copy of an existing process in operation at our company. The filter technology was obtained from a technology provider, who showed records of reliable continuous operation for many years for many commercial scale applications. After 2 weeks, the filter cloth was damaged and crystals leaked through it. The process had to be stopped and the cloth renewed. After another few weeks, the cloth was damaged again and the process had to be stopped again. Bulk chemical processes must run uninterruptedly for 46 years, because of the large investment cost, so these stoppages were really a large problem. The technology provider was consulted about the filter cloth's lifetime. After his consultation with the filter users, who were all in the food industry, it appeared that in the food industry applications the filter was cleaned every week and in the cleaning time the filter cloth was then also renewed. The real filter cloth's lifetime was unknown, but it was expected to last for a few weeks.

A team of technologists from my company was able to analyse within 1 month the cause of the rapid filter cloth wearing out and, by small adjustments of the filter equipment which took a few weeks, the filter cloth lasted several years. Everybody involved felt that this was a lucky escape.

Learning point

LP20: Purchasing a commercial scale technology proven in a different industry branch is a risk, because of differences in operation, which are not immediately obvious. Chapter 4 contains a checklist for purchasing novel processes and includes the check item: Operation details the same?

8.9 Fermentation scale-up

During my work at Shell Bioscience Laboratory at Sittingbourne, UK, a biotechnology researcher told me what had happened when his fermentation recipe developed in the laboratory was used for a large-scale production in a large-scale fermenter at a different company experienced in large-scale batch fermentation. It appeared that even after several attempts the fermentation still did not occur, although the recipe was meticulously followed. Then he visited the large-scale fermentation plant and immediately noticed that the sterilisation of the fermenter content was done by live steam injection, while he had always sterilised chemically. He then concluded that the steam injection probably stripped the ammonia from the medium. This appeared to be the case. By changing the sterilisation procedure, the problem was solved.

Learning points

LP21: Write down all relevant aspects of a fermentation recipe, and in general, write down all relevant aspects of chemistry for commercial implementation.

LP22: Have the biotechnologist or chemist researcher of the novel process available at the start-up.

Ad LP21 and 22: These learning points support the guidelines presented in Chapter 7 on start-up preparation.

8.10 Fine chemicals discovery stage

My company had acquired in the 1980s three fine chemicals business. These businesses were established companies with their own R&D, manufacturing, and marketing. I was asked to provide advice to a process research project in one of these companies. I arrived at the company and the research chemists explained their problem. They had carried out a reaction in a small tube reactor (diameter a few millimetres) filled with catalyst particles. The reaction went well. Then they decided to use a larger diameter reactor and then they could not get the reactor to run at steady state. At the same inlet temperature as the small reactor, the catalyst would flow out of tube when fed at the bottom. Then they decided to feed from the top, to avoid this catalyst flowing out. Also, for that mode of operation, they could not get a steady state. After a while, the liquid feed flow would stop.

I asked them what the heat of reaction was. They did not know, but using data from similar reactions I made an estimate and then calculated the adiabatic temperature rise of the reaction. This appeared to be over 200°C, while the reaction was supposed to be at 80°C. I explained to them that the heat of reaction was so much that the heat loss through the wall of the larger reactor was small compared to the heat of reaction, so that the reaction would run away to temperatures above the boiling point of the liquid, while for the small reactor the heat loss via the wall was significant so that the reaction did not runaway and stayed in the liquid phase. I asked them to provide me all experimental results of the project, which they did. It appeared that they also had very small experimental results from which I could draw conclusions on the reaction rate. With that data, I made a simple concept design of a multi-tubular cooled reactor with small diameter tubes. I advised them to have the reaction enthalpy experimentally determined for a better design by our process engineers at the central office. The project had lasted 2 years, of which most time was spent, to no avail, on the larger diameter test reactor.

Learning point

LP23: Involve the chemical engineer in the early part of the concept stage.

8.11 Summary of learning points from industrial cases

LP1: Product specifications are only a small part of communications with client and the client is a very important stakeholder in process innovation, as discussed in Chapter 2.

LP2: Scale-up of a single unit operation without pilot plant can be successful, if all knowledge essential for the performance is available and integrated in a validated model. This is treated in Chapter 4.

LP3: Taguchi robustness by design method, combined with a process model, is a powerful tool to create output robustness to input variation.

LP4: The so-called carbon copy of the process was in reality not a carbon copy because the local context with the re-dissolving option of the reference case was different from the new process.

Assuming that the process is conventional, while in reality it is new, is a major cause of start-up problems in general. By visiting the reference case plant, the newness was obtained. Chapter 4 contains details to determine whether a process is new.

LP5: The dynamic model appeared to be extremely useful for optimisation of the start-up procedure and for instruction of the operators.

LP6: Process start-up with operators with no or little process knowledge is asking for long start-ups and many problems. This learning point is incorporated in Chapter 7.

LP7: When purchasing process technologies, it must be checked on all details and in particular on those with scale-up effects. Section 4.3.4 treats purchasing integral processes and Section 7.3.3 treats purchasing process equipment.

LP8: Be aware of relatively large hold-ups in pilot plants which could mask problems.

LP9: Inspect all pilot plant streams on trace components build-up.

LP10: Have the process design audited and challenged by experienced people outside the design team, even if the process design is a carbon copy of an existing plant.

LP11: Appoint a manager experienced in new process start-ups as start-up leader.

LP12: Form a start-up team including engineers, process control, and process chemist.

LP13: Make a start-up plan and share it with the operating people.

LP14: Record changes and the actual process behaviour in comparison with design conditions.

LP15: For breakthrough concept designs, make the scope as wide as possible.

It may take some effort to convince managers that also wastewater treatment and other treatment parts, where conventional solutions are available, should be included in the new concept design. But the effort is justified.

LP16: Concept design using the expert system PROSYN in combination with experienced process developers can lead to novel concepts.

LP17: For a single unit operation, a pilot plant is not needed if a well-designed laboratory set-up is used.

When and when not to have a pilot plant is described in Chapter 4.

LP18: A technology provider and subcontractor with experience in a certain field are of great value.

This learning point supports the advice on the use of external parties in research and development provided in Chapters 3 and 4.

LP19: Developing a new product in combination with a new process and new market outside the core company business is a risky thing to do.

Chapter 3 contains a knowledge gap and a competence assessment to identify risks involved in a new market, a new product, and a new process.

LP20: Purchasing a commercial scale technology proven in a different industry branch is a risk, because of differences in operation, which are not immediately obvious. Chapter 4 contains a checklist for purchasing novel processes and includes the check item: Operation details the same?

LP21: Write down all relevant aspects of a fermentation recipe, and in general, write down all relevant aspects of chemistry for commercial implementation.

LP22: Have the biotechnologist or chemist researcher of the novel process available at the start-up.

These learning points support the guidelines presented in Chapter 7 on start-up preparation.

References

Ainsworth, D., 2005. Planning and Designing a Pharmaceutical Facility: A Process Designer's View. vol. 17 Pharmaceutical Technology Europe. Issue 9.

Artman, C., 2009. The Value of Information Updating in New Product Development. Springer, Berlin.

Bakker, H.L.M., et al., 2014. Management of Engineering Projects–People are Key. NAP, Nijkerk.

Bell, T.A., 2005. Challenges in the scale-up of particulate processes—an industrial perspective. Powder Technol. 150, 60–71.

Betz, F., 2011. Managing Technological Innovation: Competitive Advantage from Change, third ed. John Wiley and Sons, Hoboken, NJ.

de Bie, J.H., et al., 2001. Process for the purification of industrial waste water from a propylene oxide production process, Patent, WO 01/32561.

Bisio, A., Kabel, R.L., 1985. Scaleup of Chemical Processes: Conversion from Laboratory Scale Tests to Successful Commercial Size Design. John Wiley and Sons, New York, NY.

Bos, A.N.R., 2014. Reaction Engineering through the Funnel of Innovation, presentation, ISCRE-23 Bangkok.

Bovendeerd, B., 2012. Micro mixing in liquid-liquid systems. MSc Thesis, TU Eindhoven.

BP, 2002a. Leaps of Innovation (VAM and AVADA), BP Frontiers Magazine Issue 4. http://www.wyetec.co.uk/PDFs/BP_Frontiers_magazine_issue_4_Leaps_of_innovation.pdf.

BP, 2002b. Joffre LAO Plant, Chemicals Technology. https://www.chemicals-technology.com/projects/prairie/.

BP, 2015. BP Celebrates Start-up of Zhuhai PTA 3 Plant, Release date: 3 July 2015. https://www.bp.com/en/global/corporate/media/press-releases/bp-celebrates-start-up-of-zhuhai-pta-3-plant.html.

Charlotte, W.C., Watts, P., 2016. Micro Reaction Technology in Organic Synthesis. CRC Press, p. 453.

Dal Pont, J.P. (Ed.), 2011a. Process Engineering and Industrial Management. John Wiley and Sons, Somerset, NJ.

Dal Pont, J.-P. (Ed.), 2011b. Process Engineering and Industrial Management. John Wiley and Sons, Somerset, NJ.

van den Akker, H., Mudde, R.F., 2014. Transport Phenomena—The Art of Balancing. Delft Academic Press, Delft.

van den Akker, H.E.A., 2010. Toward a truly multiscale computational strategy for simulating turbulent two-phase flow processes. Ind. Eng. Chem. Res. 49 (21), 10780–10797.

Douglas, J.M., 1988. Conceptual Design of Chemical Processes. McGraw-Hill, New York, NY.

Dow, 1994a, AIChE. Dow's Fire & Explosion Index Hazard Classification Guide, seventh ed.

Dow, 1994b, AIChE. Dow's Chemical Exposure Index Guide, second ed.

Euzen, J.-P., 1993. Scale-Up Methodology for Chemical Processes. IFP, France.

Fink, H., Hamp, M.J., 2000. Designing and Constructing Microplants Microreaction Technology. chapter in Industrial Prospects: IMRET 3, In: Ehrfeld, W. (Ed.), Proceedings of the 3rd Int. Micro Reaction Technology. Springer, Berlin.

Fulks, B.D., 1982. Planning and organising for less troublesome plant start-ups. Chem. Eng., 96–106.

Gierman, H., 1988. Design of laboratory hydrotreating reactors, scaling down of trickle-flow reactors. Appl. Catal. 43, 277–286.

Gonzales, J., 2001. Fine Chemical Carbonyl Process, Oral Presentation at AIChE Spring Meeting, Houston, USA.

de Haan, A.B., Bosch, H., 2013. Industrial Separation Processes. De Gruyter, Berlin.

Harmsen, G.J., 1994. Kwaliteit tegen lage kosten. Robuuste processen door parametrisch ontwerpen. NPT procestechnologie, pp. 19–24.

Harmsen, G.J., 1996a. Product quality at lowest cost: robust processes by parameter and tolerance design using integration tools. In: Fransoo, J.C., Rutten, W.G.M.M. (Eds.), Proceedings of the Second International Conference on Computer Integrated Manufacturing in the Process Industries, Eindhoven, June 3–4.In: 1996, pp. 293–297.

Harmsen, G.J., 1996b. At the touch of a key. Chem. Proc. Technol. Int. 1996 (8), 143–146.

Harmsen, G.J., 1996c. Perfecting the parameters. Chem. Proc. Technol. Int. 1996 (8), 148–155.

Harmsen, G.J., 1996d. Kritische succesfactoren bij ontwerpen en opstarten van chemische fabrieken. NPT procestechnologie, pp. 15–17.

Harmsen, G.J., 2004. Industrial best practices of conceptual process design. Chem. Eng. Process. 43 (5), 671–675.

Harmsen, G.J., 2007. Reactive distillation: the frontrunner of industrial process intensification: a full review of commercial applications, research, scale-up, design and operation. Chem. Eng. Process. Process Intensification 46 (9), 774–780.

Harmsen, G.J., van Eck, D., 2004. Combining stage/gate and concurrent engineering in industrial R&D: shown in a reversible reactive extraction project, Oral presentation, Abstract and paper, in Congress Manuscripts 7th World Congress Chemical Engineering, Glasgow, 10 14 July 2005, Cdrom, IChemE.

Harmsen, G.J., Rots, A.W.T., 2009. Process for the preparation of alkylene glycol. Patent EC C07C29/12.

Harmsen, J., 2010a. Process intensification: its drivers and hurdles for commercial implementation. Chem. Eng. Process. 49, 70–73.

Harmsen, J., 2010b. In: Powell, J.B. (Ed.), Sustainable Development in the Process Industries Cases and Impact. John Wiley & Sons, Hoboken.

Harmsen, J., 2013a. Economics and environmental impact of process intensification: an assessment for the petrochemical, pharmaceutical and fine chemicals industries. In: Boodhoo, K.V.K., Harvey, A.P. (Eds.), Process Intensification for Green Chemistry. John Wiley and Sons, Chichester (Chapter 14, ISBN: 978047097267).

Harmsen, J., 2013b. Implementation of process intensification in industry. In: Boodhoo, K.V.K., Harvey, A.P. (Eds.), Process Intensification for Green Chemistry, Ch. 16. John Wiley and Sons, Chichester.

Harmsen, J., 2014. Novel sustainable industrial processes: from idea to commercial scale implementation. Green Process. Synth. 3 (3), 189–193.

Harmsen, J., 2018. In: de Haan André, B., Swinkels, P.L.J. (Eds.), Product and Process Design Driving Innovation. De Gruyter, Berlin.

Harmsen, J., Hinderink, A.P., 2000. Industrially applied process synthesis method creates synergy between economy and sustainability. In: Malone, M.M., et al., (Eds.), Fifth International Conference of Computer-Aided Process Design, Breckenridge, CO, USA July 19–24, 1999, AIChE Symposium Series No. 323.In: 96, pp. 364–366.

Harmsen, J., Jonker, G., 2012. Engineering for Sustainability: A practical Guide for Sustainable Design. Elsevier, Amsterdam.

van Helvoort, T., van Veen, R., Senden, M., 2014. Gas to Liquids—Historical Development of GTL Technology in Shell. Shell Global Solutions, Amsterdam.

Hendersen, R., 1985, Achieving quality during plant startup, Quality Prog., May 36-40.

Hoyle, W., 2002. Pilot Plants and Scale-Up of Chemical Processes. Royal Society of Chemistry, London.

Hulshof, L.A., 2013. Right First Time in Fine-chemical Process Scale-up: Avoiding Scale-up Problems: The Key to Rapid Success. Update LLP, Mayfield, UK.

ICIS, 2002. Carilon Case History, Website ICIS News. February 19.
ISO, 2017, Environmental Management—Life Cycle Assessment—Principles and Framework, sourced 11th of November 2017, https://www.google.nl/search?q=ISO%2C+2017%2C+Environmental+management+%E2%80%93+Life+cycle+assessment+%E2%80%93+Principles+and+framework&rlz=1C1CHFX_nlNL484NL484&oq=ISO%2C+2017%2C+Environmental+management+%E2%80%93+Life+cycle+assessment+%E2%80%93+Principles+and+framework&aqs=chrome..69i57.4301j0j7&sourceid=chrome&ie=UTF-8
Jain, R.K., Triandis, H.C., Weick, C.W., 2010. Managing Research, Development, and Innovation, third ed. John Wiley and Sons, Hoboken, NJ.
Jones, J.H., 2000. The Cativa™ process for the manufacture of acetic acid iridium catalyst improves productivity in an established industrial process. Platinum Metals Rev. 44 (3), 94. sourced 15 Sept. 2018, https://www.technology.matthey.com/article/44/3/94-105/.
Kane, A., 2016. Phase-appropriate formulation and process design. Pharmaceut. Technol. 40 (1), 42–45.
Kleiner, M., 2011. Smart Production by Modularisation, Oral presentation at ECCE8, Berlin.
Lager, T., 2010. Managing Process Innovation. Imperial College Press, London.
Lager, T., 2012. Startup of New Plants and Process Technology in the Process Industries: Organizing for an Extreme Event. http://www.businesschemistry.org/article/?article5148/. (Accessed 25 January 2013).
Levenspiel, O., 1999. Chemical Reaction Engineering, third ed. John Wiley and Sons, New York, NY.
Levin, M., 2006. Pharmaceutical Process Scale-up, second ed. CRC Press, Boca Raton.
Marton, A., 2011. Getting Off on the Right Foot—Innovation projects, Independent Project Analysis March Newsletter. sourced 7 November 2018, https://www.ipaglobal.com/images/Newsletter_PDFs/IPA-Newsletter-2011-Q1-Volume-3-Issue-1.pdf.
McCoy, M., 2008. Kureha's Gamble. Chem. Eng. News 86 (17), 28–29.
Merrow, E.W., 1988. Estimating startup times for solids-processing plants. Chem. Eng. 24, 89–92.
Merrow, E.W., 1991. Commercialising new technologies in the chemical process industries. In: Independent Project Analysis Report, Referred to in Harmsen, G.J., Kritische succesfactoren bij het ontwerpen en opstarten van chemische fabrieken, NPT June 1996.
Merrow, E.W., 2011. Industrial Megaprojects, Concepts, Strategies, and Practices for Success. John Wiley and Sons, New York, NY.
Omnexus, 2002. Shell donates Carilon patents to SRI International, website Omnexus, July 2.
Raimundo, P.M., 2015. Analysis and modelization of local hydrodynamics in bubble columns. Chemical and Process Engineering, Université Grenoble Alpes, 2015. Sourced 29 November 2018. https://tel.archives-ouvertes.fr/tel-01267349/document.
Ryan, C.T., 1972. Managing the process start-up. Chem. Eng. Progr. 68 (12), 65–71.
Schnider, C., 2018. Lonza, From Batch to continuous Fluorination, Presentation at Process Intensification Network, Netherlands meeting, at DIFFER, Eindhoven, 27th of June.
Schwalbe, T., 2002. Microstructured Reactor Systems. CHIMIA 2002, 56, 11, pp. 636–646.
Schwalbe, T., et al., 2002. Chemical synthesis in microreactors, microstructured reactor systems. CHIMIA 56 (11), 636–646.
Seider, W.D., Seader, J.D., Lewin, D.R., 2010. Process Design Principles: Synthesis, Analysis, and Evaluation, third ed. John Wiley and Sons, New York, NY, Hoboken, NJ.
Siirola, J.J., 1995. An Industrial Perspective on Process Synthesis in: Fourth International Conference on Foundations CAPE. In: AIChE Symp. Series No. 304. 91, pp. 222–233.
UN, 2015. Transforming Our World: The 2030 Agenda for Sustainable Development, Resolution Adopted by the General Assembly on 25 September 2015. Resourced 11th of September 2018, https://sustainabledevelopment.un.org/post2015/transformingourworld.

Verloop, J., 2004. Insight in innovation, managing innovation by understanding the laws of innovation, Shell Global Solutions. Elsevier, Amsterdam.
Verkerk, M.J., et al., 2017. Philosophy of Technology, Routledge, New York.
Verver, A., 2018. Innovation Stages in Food Industries. Personal communication Harmsen, 12 November 2018.
Vogel, A.J., 1974. Guidelines for Process Scale-Up Final Manuscript. 76th AIChE NATL MEET, Tulsa, PAP N12A, 21 pages.
Vogel, G.H., 2005. Process Development: The Initial Idea to the Chemical Production Plant. Wiley-VCH, Weinheim.
Wesselingh, J.A., Krishna, R., 1990. Mass Transfer. Ellis Horwood Ltd, London.
Westerterp, K.R., van Swaaij, W.P.M., Beenackers, A.A.C.M., 1984. Chemical Reactor Design and Operation. John Wiley & Sons, Hoboken.
Wiles, C., Watts, P., 2016. Micro Reaction Technology in Organic Synthesis. CRC Press.
Zlokarnik, M., 2002. Scale-up in Chemical Engineering, second ed. John Wiley & Sons, Somerset.
Zwietering, T.N., 1984. A backmixing model describing micromixing in single-phase continuous-flow systems. Chem. Eng. Sci. 39 (12), 1765–1778.

Index

Note: Page numbers followed by *f* indicate figures and *t* indicate tables.

T

U

V

W

Printed in the United States
By Bookmasters